VOLUME NINETY THREE

ADVANCES IN MARINE BIOLOGY

ADVANCES IN MARINE BIOLOGY

VOLUME NINETY THREE

ADVANCES IN
MARINE BIOLOGY

Edited by

CHARLES SHEPPARD

School of Life Sciences,
The University of Warwick,
United Kingdom

ELSEVIER

ACADEMIC PRESS

An imprint of Elsevier

Academic Press is an imprint of Elsevier
125 London Wall, London, EC2Y 5AS, United Kingdom
The Boulevard, Langford Lane, Kidlington, Oxford OX5 1GB, United Kingdom
525 B Street, Suite 1650, San Diego, CA 92101, United States
50 Hampshire Street, 5th Floor, Cambridge, MA 02139, United States

First edition 2022

ISBN: 978-0-323-98589-5
ISSN: 0065-2881

For information on all Academic Press publications
visit our website at https://www.elsevier.com/books-and-journals

Publisher: Zoe Kruze
Acquisitions Editor: Jason Mitchell
Developmental Editor: Cindy Angelita P. Gardose
Production Project Manager: Vijayaraj Purushothaman
Cover Designer: Matthew Limbert

Typeset by STRAIVE, India

Contributors to Volume 93

Octavio Aburto-Oropeza
Scripps Institution of Oceanography, UC San Diego, San Diego, CA, United States

Ward Appeltans
Intergovernmental Oceanographic Commission of UNESCO, Oostende, Belgium

Jorge Assis
Centre of Marine Sciences, University of Algarve, Faro, Portugal

Jiayu Bai
School of law, Nankai University, Tianjin, China

Lisa T. Ballance
Marine Mammal Institute, Oregon State University, Newport, OR, United States

Philippe Cury
MARBEC, IRD, Univ. Montpellier, Brussels, Belgium

Carlos Duarte
King Abdullah University of Science and Technology (KAUST), Red Sea Research Center (RSRC) and Computational Bioscience Research Center (CBRC), Thuwal, Kingdom of Saudi Arabia

Fabio Favoretto
Autonomous University of Baja California Sur, La Paz, Baja California Sur, Mexico

Lisa A. Hynes
Nekton Foundation, Begbroke Science Park, Oxford, United Kingdom

Joy A. Kumagai
Senckenberg Biodiversity and Climate Research Institute, Frankfurt am Main, Germany

Catherine E. Lovelock
School of Biological Sciences, The University of Queensland, St Lucia, QLD, Australia

Patricia Miloslavich
Scientific Committee on Oceanic Research (SCOR), College of Earth, Ocean and Environment, University of Delaware; Departamento de Estudios Ambientales, Universidad Simón Bolívar, Venezuela & Scientific Committee for Oceanic Research (SCOR), Newark, DE, United States

Aidin Niamir
Senckenberg Biodiversity and Climate Research Institute, Frankfurt am Main, Germany

Bethan C. O'Leary
Centre for Ecology & Conservation, College of Life and Environmental Sciences, University of Exeter, Penryn; Department of Environment and Geography, University of York, York, United Kingdom

David Obura
CORDIO East Africa, Mombasa, Kenya

Eva Ramirez-Llodra
REV Ocean, Lysaker, Norway; Nekton Foundation, Begbroke Science Park, Oxford, United Kingdom

Gabriel Reygondeau
Yale Center for Biodiversity Movement and Global Change, Yale University, New Haven, CT, United States; Nippon Foundation-Nereus Program, Institute for the Oceans and Fisheries, University of British Columbia, Vancouver, BC, Canada

Fabian Ritter
M.E.E.R. e.V., Berlin, Germany

Callum Roberts
Centre for Ecology & Conservation, College of Life and Environmental Sciences, University of Exeter, Penryn, United Kingdom

Alex D. Rogers
REV Ocean, Lysaker, Norway; Nekton Foundation, Begbroke Science Park, Oxford, United Kingdom

Yvonne Sadovy
School of Biological Sciences, Swire Institute of Marine Science, The University of Hong Kong, Hong Kong

Oliver Steeds
Nekton Foundation, Begbroke Science Park, Oxford, United Kingdom

Tracey Sutton
Nova Southeastern University, Halmos College of Natural Sciences and Oceanography, Dania Beach, FL, United States

Derek P. Tittensor
Department of Biology, Dalhousie University, Halifax, NS, Canada

Enriqueta Velarde
Instituto de Ciencias Marinas y Pesquerías, Universidad Veracruzana, Veracruz, Mexico

Lucy Woodall
Nekton Foundation, Begbroke Science Park; Department of Zoology, University of Oxford, Oxford, United Kingdom

Kailei Zhu
School of International Affairs and public Administration, Ocean University of China, Qingdao, Shandong, China

Contents

Contributors to Volume 93 *v*
Series Contents for Last Fifteen Years *ix*

1. Opinion: Marine mammal conservation in the 21st century: A plea for a paradigm shift towards mindful conservation **1**
Fabian Ritter

1. Preface 4
2. Change 6
3. Three ways of marine (mammal) conservation 8
4. Mindful conservation and the commons 14
5. The personal (but not necessarily private) dimension 15
6. Conclusion 17

Acknowledgements 18
References 18

2. Discovering marine biodiversity in the 21st century **23**
Alex D. Rogers, Ward Appeltans, Jorge Assis, Lisa T. Ballance, Philippe Cury, Carlos Duarte, Fabio Favoretto, Lisa A. Hynes, Joy A. Kumagai, Catherine E. Lovelock, Patricia Miloslavich, Aidin Niamir, David Obura, Bethan C. O'Leary, Eva Ramirez-Llodra, Gabriel Reygondeau, Callum Roberts, Yvonne Sadovy, Oliver Steeds, Tracey Sutton, Derek P. Tittensor, Enriqueta Velarde, Lucy Woodall, and Octavio Aburto-Oropeza

1. Marine biodiversity: What is known? 24
2. International legislation regarding marine biodiversity and its implementation 44
3. How marine biodiversity is studied today 47
4. Conclusion: Integrated approaches 93

Acknowledgements 96
References 96

3. Review study on governance and international law for coastal and marine ecosystems in response to climate change: Social science perspective **117**

Kailei Zhu and Jiayu Bai

1. Introduction 118
2. Overview 119
3. Theoretical research on coastal and marine ecosystems in response to climate change 128
4. Methodological studies on coastal and marine ecosystems in response to climate change 131
5. Empirical studies on coastal and marine ecosystems in response to climate change 133
6. Discussion 137
7. Conclusion 140

Acknowledgement 140

References 140

Series Contents for Last Fifteen Years*,†

Volume 44, 2003.

Hirst, A. G., Roff, J. C. and Lampitt, R. S. A synthesis of growth rates in epipelagic invertebrate zooplankton. pp. 3–142.

Boletzky, S. von. Biology of early life stages in cephalopod molluscs. pp. 143–203.

Pittman, S. J. and McAlpine, C. A. Movements of marine fish and decapod crustaceans: process, theory and application. pp. 205–294.

Cutts, C. J. Culture of harpacticoid copepods: potential as live feed for rearing marine fish. pp. 295–315.

Volume 45, 2003.

Cumulative Taxonomic and Subject Index.

Volume 46, 2003.

Gooday, A. J. Benthic foraminifera (Protista) as tools in deep-water palaeoceanography: environmental influences on faunal characteristics. pp. 1–90.

Subramoniam, T. and Gunamalai, V. Breeding biology of the intertidal sand crab, Emerita (Decapoda: Anomura). pp. 91–182.

Coles, S. L. and Brown, B. E. Coral bleaching—capacity for acclimatization and adaptation. pp. 183–223.

Dalsgaard J., St. John M., Kattner G., Müller-Navarra D. and Hagen W. Fatty acid trophic markers in the pelagic marine environment. pp. 225–340.

Volume 47, 2004.

Southward, A. J., Langmead, O., Hardman-Mountford, N. J., Aiken, J., Boalch, G. T., Dando, P. R., Genner, M. J., Joint, I., Kendall, M. A., Halliday, N. C., Harris, R. P., Leaper, R., Mieszkowska, N., Pingree, R. D., Richardson, A. J., Sims, D.W., Smith, T., Walne, A. W. and Hawkins, S. J. Long-term oceanographic and ecological research in the western English Channel. pp. 1–105.

Queiroga, H. and Blanton, J. Interactions between behaviour and physical forcing in the control of horizontal transport of decapod crustacean larvae. pp. 107–214.

*The full list of contents for volumes 1–37 can be found in volume 38

†The full list of contents for volumes 38–43 can be found in volume 78

Braithwaite, R. A. and McEvoy, L. A. Marine biofouling on fish farms and its remediation. pp. 215–252.

Frangoulis, C., Christou, E. D. and Hecq, J. H. Comparison of marine copepod outfluxes: nature, rate, fate and role in the carbon and nitrogen cycles. pp. 253–309.

Volume 48, 2005.

Canfield, D. E., Kristensen, E. and Thamdrup, B. Aquatic Geomicrobiology. pp. 1–599.

Volume 49, 2005.

Bell, J. D., Rothlisberg, P. C., Munro, J. L., Loneragan, N. R., Nash, W. J., Ward, R. D. and Andrew, N. L. Restocking and stock enhancement of marine invertebrate fisheries. pp. 1–358.

Volume 50, 2006.

Lewis, J. B. Biology and ecology of the hydrocoral Millepora on coral reefs. pp. 1–55.

Harborne, A. R., Mumby, P. J., Micheli, F., Perry, C. T., Dahlgren, C. P., Holmes, K. E., and Brumbaugh, D. R. The functional value of Caribbean coral reef, seagrass and mangrove habitats to ecosystem processes. pp. 57–189.

Collins, M. A. and Rodhouse, P. G. K. Southern ocean cephalopods. pp. 191–265.

Tarasov, V. G. Effects of shallow-water hydrothermal venting on biological communities of coastal marine ecosystems of the western Pacific. pp. 267–410.

Volume 51, 2006.

Elena Guijarro Garcia. The fishery for Iceland scallop (Chlamys islandica) in the Northeast Atlantic. pp. 1–55.

Jeffrey, M. Leis. Are larvae of demersal fishes plankton or nekton? pp. 57–141.

John C. Montgomery, Andrew Jeffs, Stephen D. Simpson, Mark Meekan and Chris Tindle. Sound as an orientation cue for the pelagic larvae of reef fishes and decapod crustaceans. pp. 143–196.

Carolin E. Arndt and Kerrie M. Swadling. Crustacea in Arctic and Antarctic sea ice: Distribution, diet and life history strategies. pp. 197–315.

Volume 52, 2007.
Leys, S. P., Mackie, G. O. and Reiswig, H. M. The Biology of Glass Sponges. pp. 1–145.
Garcia E. G. The Northern Shrimp (Pandalus borealis) Offshore Fishery in the Northeast Atlantic. pp. 147–266.
Fraser K. P. P. and Rogers A. D. Protein Metabolism in Marine Animals: The Underlying Mechanism of Growth. pp. 267–362.

Volume 53, 2008.
Dustin J. Marshall and Michael J. Keough. The Evolutionary Ecology of Offspring Size in Marine Invertebrates. pp. 1–60.
Kerry A. Naish, Joseph E. Taylor III, Phillip S. Levin, Thomas P. Quinn, James R. Winton, Daniel Huppert, and Ray Hilborn. An Evaluation of the Effects of Conservation and Fishery Enhancement Hatcheries on Wild Populations of Salmon. pp. 61–194.
Shannon Gowans, Bernd Würsig, and Leszek Karczmarski. The Social Structure and Strategies of Delphinids: Predictions Based on an Ecological Framework. pp. 195–294.

Volume 54, 2008.
Bridget S. Green. Maternal Effects in Fish Populations. pp. 1–105.
Victoria J. Wearmouth and David W. Sims. Sexual Segregation in Marine Fish, Reptiles, Birds and Mammals: Behaviour Patterns, Mechanisms and Conservation Implications. pp. 107–170.
David W. Sims. Sieving a Living: A Review of the Biology, Ecology and Conservation Status of the Plankton-Feeding Basking Shark *Cetorhinus Maximus*. pp. 171–220.
Charles H. Peterson, Kenneth W. Able, Christin Frieswyk DeJong, Michael F. Piehler, Charles A. Simenstad, and Joy B. Zedler. Practical Proxies for Tidal Marsh Ecosystem Services: Application to Injury and Restoration. pp. 221–266.

Volume 55, 2008.
Annie Mercier and Jean-Francois Hamel. Introduction. pp. 1–6.
Annie Mercier and Jean-Francois Hamel. Gametogenesis. pp. 7–72.
Annie Mercier and Jean-Francois Hamel. Spawning. pp. 73–168.
Annie Mercier and Jean-Francois Hamel. Discussion. pp. 169–194.

Volume 56, 2009.

Philip C. Reid, Astrid C. Fischer, Emily Lewis-Brown, Michael P. Meredith, Mike Sparrow, Andreas J. Andersson, Avan Antia, Nicholas R. Bates, Ulrich Bathmann, Gregory Beaugrand, Holger Brix, Stephen Dye, Martin Edwards, Tore Furevik, Reidun Gangst, Hjalmar Hatun, Russell R. Hopcroft, Mike Kendall, Sabine Kasten, Ralph Keeling, Corinne Le Quere, Fred T. Mackenzie, Gill Malin, Cecilie Mauritzen, Jon Olafsson, Charlie Paull, Eric Rignot, Koji Shimada, Meike Vogt, Craig Wallace, Zhaomin Wang and Richard Washington. Impacts of the Oceans on Climate Change. pp. 1–150.

Elvira S. Poloczanska, Colin J. Limpus and Graeme C. Hays. Vulnerability of Marine Turtles to Climate Change. pp. 151–212.

Nova Mieszkowska, Martin J. Genner, Stephen J. Hawkins and David W. Sims. Effects of Climate Change and Commercial Fishing on Atlantic Cod *Gadus morhua*. pp. 213–274.

Iain C. Field, Mark G. Meekan, Rik C. Buckworth and Corey J. A. Bradshaw. Susceptibility of Sharks, Rays and Chimaeras to Global Extinction. pp. 275–364.

Milagros Penela-Arenaz, Juan Bellas and Elsa Vazquez. Effects of the *Prestige* Oil Spill on the Biota of NW Spain: 5 Years of Learning. pp. 365–396.

Volume 57, 2010.

Geraint A. Tarling, Natalie S. Ensor, Torsten Fregin, William P. Good-all-Copestake and Peter Fretwell. An Introduction to the Biology of Northern Krill (*Meganyctiphanes norvegica* Sars). pp. 1–40.

Tomaso Patarnello, Chiara Papetti and Lorenzo Zane. Genetics of Northern Krill (*Meganyctiphanes norvegica* Sars). pp. 41–58.

Geraint A. Tarling. Population Dynamics of Northern Krill (*Meganyctiphanes norvegica* Sars). pp. 59–90.

John I. Spicer and Reinhard Saborowski. Physiology and Metabolism of Northern Krill (*Meganyctiphanes norvegica* Sars). pp. 91–126.

Katrin Schmidt. Food and Feeding in Northern Krill (*Meganyctiphanes norvegica* Sars). pp. 127–172.

Friedrich Buchholz and Cornelia Buchholz. Growth and Moulting in Northern Krill (*Meganyctiphanes norvegica* Sars). pp. 173–198.

Janine Cuzin-Roudy. Reproduction in Northern Krill. pp. 199–230.

Edward Gaten, Konrad Wiese and Magnus L. Johnson. Laboratory-Based Observations of Behaviour in Northern Krill (*Meganyctiphanes norvegica* Sars). pp. 231–254.

Stein Kaartvedt. Diel Vertical Migration Behaviour of the Northern Krill (*Meganyctiphanes norvegica* Sars). pp. 255–276.
Yvan Simard and Michel Harvey. Predation on Northern Krill (*Meganyctiphanes norvegica* Sars). pp. 277–306.

Volume 58, 2010.
A. G. Glover, A. J. Gooday, D. M. Bailey, D. S. M. Billett, P. Chevaldonné, A. Colaço, J. Copley, D. Cuvelier, D. Desbruyères, V. Kalogeropoulou, M. Klages, N. Lampadariou, C. Lejeusne, N. C. Mestre, G. L. J. Paterson, T. Perez, H. Ruhl, J. Sarrazin, T. Soltwedel, E. H. Soto, S. Thatje, A. Tselepides, S. Van Gaever, and A. Vanreusel. Temporal Change in Deep-Sea Benthic Ecosystems: A Review of the Evidence From Recent Time-Series Studies. pp. 1–96.
Hilario Murua. The Biology and Fisheries of European Hake, *Merluccius merluccius*, in the North-East Atlantic. pp. 97–154.
Jacopo Aguzzi and Joan B. Company. Chronobiology of Deep-Water Decapod Crustaceans on Continental Margins. pp. 155–226.
Martin A. Collins, Paul Brickle, Judith Brown, and Mark Belchier. The Patagonian Toothfish: Biology, Ecology and Fishery. pp. 227–300.

Volume 59, 2011.
Charles W. Walker, Rebecca J. Van Beneden, Annette F. Muttray, S. Anne Böttger, Melissa L. Kelley, Abraham E. Tucker, and W. Kelley Thomas. p53 Superfamily Proteins in Marine Bivalve Cancer and Stress Biology. pp 1–36.
Martin Wahl, Veijo Jormalainen, Britas Klemens Eriksson, James A. Coyer, Markus Molis, Hendrik Schubert, Megan Dethier, Anneli Ehlers, Rolf Karez, Inken Kruse, Mark Lenz, Gareth Pearson, Sven Rohde, Sofia A. Wikström, and Jeanine L. Olsen. Stress Ecology in *Fucus*: Abiotic, Biotic and Genetic Interactions. pp. 37–106.
Steven R. Dudgeon and Janet E. Kübler. Hydrozoans and the Shape of Things to Come. pp. 107–144.
Miles Lamare, David Burritt, and Kathryn Lister. Ultraviolet Radiation and Echinoderms: Past, Present and Future Perspectives. pp. 145–187.

Volume 60, 2011.
Tatiana A. Rynearson and Brian Palenik. Learning to Read the Oceans: Genomics of Marine Phytoplankton. pp. 1–40.
Les Watling, Scott C. France, Eric Pante and Anne Simpson. Biology of Deep-Water Octocorals. pp. 41–122.

Cristián J. Monaco and Brian Helmuth. Tipping Points, Thresholds and the Keystone Role of Physiology in Marine Climate Change Research. pp. 123–160.

David A. Ritz, Alistair J. Hobday, John C. Montgomery and Ashley J.W. Ward. Social Aggregation in the Pelagic Zone with Special Reference to Fish and Invertebrates. pp. 161–228.

Volume 61, 2012.

Gert Wörheide, Martin Dohrmann, Dirk Erpenbeck, Claire Larroux, Manuel Maldonado, Oliver Voigt, Carole Borchiellini and Denis Lavrov. Deep Phylogeny and Evolution of Sponges (Phylum Porifera). pp. 1–78.

Paco Cárdenas, Thierry Pérez and Nicole Boury-Esnault. Sponge Systematics Facing New Challenges. pp. 79–210.

Klaus Rützler. The Role of Sponges in the Mesoamerican Barrier-Reef Ecosystem, Belize. pp. 211–272.

Janie Wulff. Ecological Interactions and the Distribution, Abundance, and Diversity of Sponges. pp. 273–344.

Maria J. Uriz and Xavier Turon. Sponge Ecology in the Molecular Era. pp. 345–410.

Volume 62, 2012.

Sally P. Leys and April Hill. The Physiology and Molecular Biology of Sponge Tissues. pp. 1–56.

Robert W. Thacker and Christopher J. Freeman. Sponge–Microbe Symbioses: Recent Advances and New Directions. pp. 57–112.

Manuel Maldonado, Marta Ribes and Fleur C. van Duyl. Nutrient Fluxes Through Sponges: Biology, Budgets, and Ecological Implications. pp. 113–182.

Grégory Genta-Jouve and Olivier P. Thomas. Sponge Chemical Diversity: From Biosynthetic Pathways to Ecological Roles. pp. 183–230.

Xiaohong Wang, Heinz C. Schröder, Matthias Wiens, Ute Schloßmacher and Werner E. G. Müller. Biosilica: Molecular Biology, Biochemistry and Function in Demosponges as well as its Applied Aspects for Tissue Engineering. pp. 231–272.

Klaske J. Schippers, Detmer Sipkema, Ronald Osinga, Hauke Smidt, Shirley A. Pomponi, Dirk E. Martens and René H. Wijffels. Cultivation of Sponges, Sponge Cells and Symbionts: Achievements and Future Prospects. pp. 273–338.

Volume 63, 2012.
Michael Stat, Andrew C. Baker, David G. Bourne, Adrienne M. S. Correa, Zac Forsman, Megan J. Huggett, Xavier Pochon, Derek Skillings, Robert J. Toonen, Madeleine J. H. van Oppen, and Ruth D. Gates. Molecular Delineation of Species in the Coral Holobiont. pp. 1–66.
Daniel Wagner, Daniel G. Luck, and Robert J. Toonen. The Biology and Ecology of Black Corals (Cnidaria: Anthozoa: Hexacorallia: Antipatharia). pp. 67–132.
Cathy H. Lucas, William M. Graham, and Chad Widmer. Jellyfish Life Histories: Role of Polyps in Forming and Maintaining Scyphomedusa Populations. pp. 133–196.
T. Aran Mooney, Maya Yamato, and Brian K. Branstetter. Hearing in Cetaceans: From Natural History to Experimental Biology. pp. 197–246.

Volume 64, 2013.
Dale Tshudy. Systematics and Position of *Nephrops* Among the Lobsters. pp. 1–26.
Mark P. Johnson, Colm Lordan, and Anne Marie Power. Habitat and Ecology of *Nephrops norvegicus*. pp. 27–64.
Emi Katoh, Valerio Sbragaglia, Jacopo Aguzzi, and Thomas Breithaupt. Sensory Biology and Behaviour of *Nephrops norvegicus*. pp. 65–106.
Edward Gaten, Steve Moss, and Magnus L. Johnson. The Reniform Reflecting Superposition Compound Eyes of *Nephrops norvegicus*: Optics, Susceptibility to Light-Induced Damage, Electrophysiology and a Ray Tracing Model. pp. 107–148.
Susanne P. Eriksson, Bodil Hernroth, and Susanne P. Baden. Stress Biology and Immunology in *Nephrops norvegicus*. pp. 149–200.
Adam Powell and Susanne P. Eriksson. Reproduction: Life Cycle, Larvae and Larviculture. pp. 201–246.
Anette Ungfors, Ewen Bell, Magnus L. Johnson, Daniel Cowing, Nicola C. Dobson, Ralf Bublitz, and Jane Sandell. *Nephrops* Fisheries in European Waters. pp. 247–314.

Volume 65, 2013.
Isobel S.M. Bloor, Martin J. Attrill, and Emma L. Jackson. A Review of the Factors Influencing Spawning, Early Life Stage Survival and Recruitment Variability in the Common Cuttlefish (*Sepia officinalis*). pp. 1–66.
Dianna K. Padilla and Monique M. Savedo. A Systematic Review of Phenotypic Plasticity in Marine Invertebrate and Plant Systems. pp. 67–120.

Leif K. Rasmuson. The Biology, Ecology and Fishery of the Dungeness crab, *Cancer magister*. pp. 121–174.

Volume 66, 2013.

Lisa-ann Gershwin, Anthony J. Richardson, Kenneth D. Winkel, Peter J. Fenner, John Lippmann, Russell Hore, Griselda Avila-Soria, David Brewer, Rudy J. Kloser, Andy Steven, and Scott Condie. Biology and Ecology of Irukandji Jellyfish (Cnidaria: Cubozoa). pp. 1–86.

April M. H. Blakeslee, Amy E. Fowler, and Carolyn L. Keogh. Marine Invasions and Parasite Escape: Updates and New Perspectives. pp. 87–170.

Michael P. Russell. Echinoderm Responses to Variation in Salinity. pp. 171–212.

Daniela M. Ceccarelli, A. David McKinnon, Serge Andréfouët, Valerie Allain, Jock Young, Daniel C. Gledhill, Adrian Flynn, Nicholas J. Bax, Robin Beaman, Philippe Borsa, Richard Brinkman, Rodrigo H. Bustamante, Robert Campbell, Mike Cappo, Sophie Cravatte, Stéphanie D'Agata, Catherine M. Dichmont, Piers K. Dunstan, Cécile Dupouy, Graham Edgar, Richard Farman, Miles Furnas, Claire Garrigue, Trevor Hutton, Michel Kulbicki, Yves Letourneur, Dhugal Lindsay, Christophe Menkes, David Mouillot, Valeriano Parravicini, Claude Payri, Bernard Pelletier, Bertrand Richer de Forges, Ken Ridgway, Martine Rodier, Sarah Samadi, David Schoeman, Tim Skewes, Steven Swearer, Laurent Vigliola, Laurent Wantiez, Alan Williams, Ashley Williams, and Anthony J. Richardson. The Coral Sea: Physical Environment, Ecosystem Status and Biodiversity Assets. pp. 213–290.

Volume 67, 2014.

Erica A.G. Vidal, Roger Villanueva, José P. Andrade, Ian G. Gleadall, José Iglesias, Noussithé Koueta, Carlos Rosas, Susumu Segawa, Bret Grasse, Rita M. Franco-Santos, Caroline B. Albertin, Claudia Caamal-Monsreal, Maria E. Chimal, Eric Edsinger-Gonzales, Pedro Gallardo, Charles Le Pabic, Cristina Pascual, Katina Roumbedakis, and James Wood. Cephalopod Culture: Current Status of Main Biological Models and Research Priorities. pp. 1–98.

Paul G.K. Rodhouse, Graham J. Pierce, Owen C. Nichols, Warwick H.H. Sauer, Alexander I. Arkhipkin, Vladimir V. Laptikhovsky, Marek R. Lipiński, Jorge E. Ramos, Michaël Gras, Hideaki Kidokoro, Kazuhiro Sadayasu, João Pereira, Evgenia Lefkaditou, Cristina Pita, Maria

Gasalla, Manuel Haimovici, Mitsuo Sakai, and Nicola Downey. Environmental Effects on Cephalopod Population Dynamics: Implications for Management of Fisheries. pp. 99–234.

Henk-Jan T. Hoving, José A.A. Perez, Kathrin Bolstad, Heather Braid, Aaron B. Evans, Dirk Fuchs, Heather Judkins, Jesse T. Kelly, José E.A.R. Marian, Ryuta Nakajima, Uwe Piatkowski, Amanda Reid, Michael Vecchione, and José C.C. Xavier. The Study of Deep-Sea Cephalopods. pp. 235–362.

Jean-Paul Robin, Michael Roberts, Lou Zeidberg, Isobel Bloor, Almendra Rodriguez, Felipe Briceño, Nicola Downey, Maite Mascaró, Mike Navarro, Angel Guerra, Jennifer Hofmeister, Diogo D. Barcellos, Silvia A.P. Lourenço, Clyde F.E. Roper, Natalie A. Moltschaniwskyj, Corey P. Green, and Jennifer Mather. Transitions During Cephalopod Life History: The Role of Habitat, Environment, Functional Morphology and Behaviour. pp. 363–440.

Volume 68, 2014.

Paul K.S. Shin, Siu Gin Cheung, Tsui Yun Tsang, and Ho Yin Wai. Ecology of Artificial Reefs in the Subtropics. pp. 1–64.

Hrafnkell Eiríksson. Reproductive Biology of Female Norway Lobster, *Nephrops norvegicus* (Linnaeus, 1758) Leach, in Icelandic Waters During the Period 1960–2010: Comparative Overview of Distribution Areas in the Northeast Atlantic and the Mediterranean. pp. 65–210.

Volume 69, 2014.

Ray Hilborn. Introduction to Marine Managed Areas. pp. 1–14.

Philip N. Trathan, Martin A. Collins, Susie M. Grant, Mark Belchier, David K.A. Barnes, Judith Brown, and Iain J. Staniland. The South Georgia and the South Sandwich Islands MPA: Protecting A Biodiverse Oceanic Island Chain Situated in the Flow of the Antarctic Circumpolar Current. pp. 15–78.

Richard P. Dunne, Nicholas V.C. Polunin, Peter H. Sand, and Magnus L. Johnson. The Creation of the Chagos Marine Protected Area: A Fisheries Perspective. pp. 79–128.

Michelle T. Schärer-Umpierre, Daniel Mateos-Molina, Richard Appeldoorn, Ivonne Bejarano, Edwin A. Hernández-Delgado, Richard S. Nemeth, Michael I. Nemeth, Manuel Valdés-Pizzini, and Tyler B. Smith. Marine Managed Areas and Associated Fisheries in the US Caribbean. pp. 129–152.

Alan M. Friedlander, Kostantinos A. Stamoulis, John N. Kittinger, Jeffrey C. Drazen, and Brian N. Tissot. Understanding the Scale of

Marine Protection in Hawai'i: From Community-Based Management to the Remote Northwestern Hawaiian Islands. pp. 153–204.
Louis W. Botsford, J. Wilson White, Mark H. Carr, and Jennifer E. Caselle. Marine Protected Area Networks in California, USA. pp. 205–252.
Bob Kearney and Graham Farebrother. Inadequate Evaluation and Management of Threats in Australia's Marine Parks, Including the Great Barrier Reef, Misdirect Marine Conservation. pp. 253–288.
Randi Rotjan, Regen Jamieson, Ben Carr, Les Kaufman, Sangeeta Mangubhai, David Obura, Ray Pierce, Betarim Rimon, Bud Ris, Stuart Sandin, Peter Shelley, U. Rashid Sumaila, Sue Taei, Heather Tausig, Tukabu Teroroko, Simon Thorrold, Brooke Wikgren, Teuea Toatu, and Greg Stone. Establishment, Management, and Maintenance of the Phoenix Islands Protected Area. pp. 289–324.
Alex J. Caveen, Clare Fitzsimmons, Margherita Pieraccini, Euan Dunn, Christopher J. Sweeting, Magnus L. Johnson, Helen Bloomfield, Estelle V. Jones, Paula Lightfoot, Tim S. Gray, Selina M. Stead, and Nicholas V. C. Polunin. Diverging Strategies to Planning an Ecologically Coherent Network of MPAs in the North Sea: The Roles of Advocacy, Evidence and Pragmatism in the Face of Uncertaintya. pp. 325–370.
Carlo Pipitone, Fabio Badalamenti, Tomás Vega Fernández, and Giovanni D'Anna. Spatial Management of Fisheries in the Mediterranean Sea: Problematic Issues and a Few Success Stories. pp. 371–402.

Volume 70, 2015.
Alex D. Rogers, Christopher Yesson, and Pippa Gravestock. A Biophysical and Economic Profile of South Georgia and the South Sandwich Islands as Potential Large-Scale Antarctic Protected Areas. pp. 1–286.

Volume 71, 2015.
Ricardo Calado and Miguel Costa Leal. Trophic Ecology of Benthic Marine Invertebrates with Bi-Phasic Life Cycles: What Are We Still Missing? pp. 1–70.
Jesse M.A. van der Grient and Alex D. Rogers. Body Size Versus Depth: Regional and Taxonomical Variation in Deep-Sea Meio- and Macrofaunal Organisms. pp. 71–108.
Lorena Basso, Maite Vázquez-Luis, José R. García-March, Salud Deudero, Elvira Alvarez, Nardo Vicente, Carlos M. Duarte, and Iris E. Hendriks. The Pen Shell, *Pinna nobilis*: A Review of Population Status and Recommended Research Priorities in the Mediterranean Sea. pp. 109–160.

Volume 72, 2015.
Thomas A. Jefferson and Barbara E. Curry. Humpback Dolphins: A Brief Introduction to the Genus *Sousa*. pp. 1–16.
Sarah Piwetz, David Lundquist, and Bernd Würsig. Humpback Dolphin (Genus *Sousa*) Behavioural Responses to Human Activities. pp. 17–46.
Tim Collins. Re-assessment of the Conservation Status of the Atlantic Humpback Dolphin, *Sousa teuszii* (Kükenthal, 1892) Using the IUCN Red List Criteria. pp. 47–78.
Caroline R. Weir and Tim Collins. A Review of the Geographical Distribution and Habitat of the Atlantic Humpback Dolphin (*Sousa teuszii*). pp. 79–118.
Gill T Braulik, Ken Findlay, Salvatore Cerchio, and Robert Baldwin. Assessment of the Conservation Status of the Indian Ocean Humpback Dolphin (*Sousa plumbea*) Using the IUCN Red List Criteria. pp. 119–142.
Stephanie Plön, Victor G. Cockcroft, and William P. Froneman. The Natural History and Conservation of Indian Ocean Humpback Dolphins (*Sousa plumbea*) in South African Waters. pp. 143–162.
Salvatore Cerchio, Norbert Andrianarivelo, and Boris Andrianantenaina. Ecology and Conservation Status of Indian Ocean Humpback Dolphins (*Sousa plumbea*) in Madagascar. pp. 163–200.
Muhammad Shoaib Kiani and Koen Van Waerebeek. A Review of the Status of the Indian Ocean Humpback Dolphin *Sousa plumbea* in Pakistan. pp. 201–228.
Dipani Sutaria, Divya Panicker, Ketki Jog, Mihir Sule, Rahul Muralidharan, and Isha Bopardikar. Humpback Dolphins (Genus *Sousa*) in India: An Overview of Status and Conservation Issues. pp. 229–256.

Volume 73, 2016.
Thomas A. Jefferson and Brian D. Smith. Re-assessment of the Conservation Status of the Indo-Pacific Humpback Dolphin (*Sousa chinensis*) Using the IUCN Red List Criteria. pp. 1–26.
Leszek Karczmarski, Shiang-Lin Huang, Carmen K. M. Or, Duan Gui, Stephen C. Y. Chan, Wenzhi Lin, Lindsay Porter, Wai-Ho Wong, Ruiqiang Zheng, Yuen-Wa Ho, Scott Y. S. Chui, Angelico Jose C. Tiongson, Yaqian Mo, Wei-Lun Chang, John H. W. Kwok, Ricky W. K. Tang, Andy T. L. Lee, Sze-Wing Yiu, Mark Keith, Glenn Gailey, and Yuping Wu. Humpback Dolphins in Hong Kong and the Pearl River Delta: Status, Threats and Conservation Challenges. pp. 27–64.

Bernd Würsig, E.C.M. Parsons, Sarah Piwetz, and Lindsay Porter. The Behavioural Ecology of Indo-Pacific Humpback Dolphins in Hong Kong. pp. 65–90.

John Y. Wang, Kimberly N. Riehl, Michelle N. Klein, Shiva Javdan, Jordan M. Hoffman, Sarah Z. Dungan, Lauren E. Dares, and Claryana Araújo-Wang. Biology and Conservation of the Taiwanese Humpback Dolphin, *Sousa chinensis taiwanensis*. pp. 91–118.

Bingyao Chen, Xinrong Xu, Thomas A. Jefferson, Paula A. Olson, Qiurong Qin, Hongke Zhang, Liwen He, and Guang Yang. Conservation Status of the Indo-Pacific Humpback Dolphin (*Sousa chinensis*) in the Northern Beibu Gulf, China. pp. 119–140.

Gianna Minton, Anna Norliza Zulkifli Poh, Cindy Peter, Lindsay Porter, and Danielle Kreb. Indo-Pacific Humpback Dolphins in Borneo: A Review of Current Knowledge with Emphasis on Sarawak. pp. 141–156.

Guido J. Parra and Daniele Cagnazzi. Conservation Status of the Australian Humpback Dolphin (*Sousa sahulensis*) Using the IUCN Red List Criteria. pp. 157–192.

Daniella M. Hanf, Tim Hunt, and Guido J. Parra. Humpback Dolphins of Western Australia: A Review of Current Knowledge and Recommendations for Future Management. pp. 193–218.

Isabel Beasley, Maria Jedensjö, Gede Mahendra Wijaya, Jim Anamiato, Benjamin Kahn, and Danielle Kreb. Observations on Australian Humpback Dolphins (*Sousa sahulensis*) in Waters of the Pacific Islands and New Guinea. pp. 219–272.

Alexander M. Brown, Lars Bejder, Guido J. Parra, Daniele Cagnazzi, Tim Hunt, Jennifer L. Smith, and Simon J. Allen. Sexual Dimorphism and Geographic Variation in Dorsal Fin Features of Australian Humpback Dolphins, *Sousa sahulensis*. pp. 273–314.

Volume 74, 2016.

J. Salinger, A.J. Hobday, R.J. Matear, T.J. O'Kane, J.S. Risbey, P. Dunstan, J.P. Eveson, E.A. Fulton, M. Feng, É.E. Plagányi, E.S. Poloczanska, A.G. Marshall, and P.A. Thompson. Decadal-Scale Forecasting of Climate Drivers for Marine Applications. pp. 1–68.

S.A. Foo and M. Byrne. Acclimatization and Adaptive Capacity of Marine Species in a Changing Ocean. pp. 69–116.

N.D. Gallo and L.A. Levin. Fish Ecology and Evolution in the World's Oxygen Minimum Zones and Implications of Ocean Deoxygenation. pp. 117–198.

R.J. Olson, J.W. Young, F. Ménard, M. Potier, V. Allain, N. Goñi, J.M. Logan, and F. Galván-Magaña. Bioenergetics, Trophic Ecology, and Niche Separation of Tunas. pp. 199–344.

Volume 75, 2016.

G. Notarbartolo di Sciara. Marine Mammals in the Mediterranean Sea: An Overview. pp. 1–36.

L. Rendell and A. Frantzis. Mediterranean Sperm Whales, *Physeter macrocephalus*: The Precarious State of a Lost Tribe. pp. 37–74.

G. Notarbartolo di Sciara, M. Castellote, J.-N. Druon, and S. Panigada. Fin Whales, *Balaenoptera physalus*: At Home in a Changing Mediterranean Sea?. pp. 75–102.

M. Podestà, A. Azzellino, A. Cañadas, A. Frantzis, A. Moulins, M. Rosso, P. Tepsich, and C. Lanfredi. Cuvier's Beaked Whale, *Ziphius cavirostris*, Distribution and Occurrence in the Mediterranean Sea: High-Use Areas and Conservation Threats. pp. 103–140.

R. Esteban, P. Verborgh, P. Gauffier, D. Alarcón, J.M. Salazar-Sierra, J. Giménez, A.D. Foote, and R. de Stephanis. Conservation Status of Killer Whales, *Orcinus orca*, in the Strait of Gibraltar. pp. 141–172.

P. Verborgh, P. Gauffier, R. Esteban, J. Giménez, A. Cañadas, J.M. Salazar-Sierra, and R. de Stephanis. Conservation Status of Long-Finned Pilot Whales, *Globicephala melas*, in the Mediterranean Sea. pp. 173–204.

A. Azzellino, S. Airoldi, S. Gaspari, C. Lanfredi, A. Moulins, M. Podestà, M. Rosso, and P. Tepsich. Risso's Dolphin, *Grampus griseus*, in the Western Ligurian Sea: Trends in Population Size and Habitat Use. pp. 205–232.

D. Kerem, O. Goffman, M. Elasar, N. Hadar, A. Scheinin, and T. Lewis. The Rough-Toothed Dolphin, *Steno bredanensis*, in the Eastern Mediterranean Sea: A Relict Population?. pp. 233–258.

J. Gonzalvo, G. Lauriano, P.S. Hammond, K.A. Viaud-Martinez, M.C. Fossi, A. Natoli, and L. Marsili. The Gulf of Ambracia's Common Bottlenose Dolphins, *Tursiops truncatus*: A Highly Dense and yet Threatened Population. pp. 259–296.

G. Bearzi, S. Bonizzoni, N.L. Santostasi, N.B. Furey, L. Eddy, V.D. Valavanis, and O. Gimenez. Dolphins in a Scaled-Down Mediterranean: The Gulf of Corinth's Odontocetes. pp. 297–332.

M.C. Fontaine. Harbour Porpoises, *Phocoena phocoena*, in the Mediterranean Sea and Adjacent Regions: Biogeographic Relicts of the Last Glacial Period. pp. 333–358.

G. Notarbartolo di Sciara and S. Kotomatas. Are Mediterranean Monk Seals, *Monachus monachus*, Being Left to Save Themselves from Extinction?. pp. 359–386.
T. Scovazzi. The International Legal Framework for Marine Mammal Conservation in the Mediterranean Sea. pp. 387–416.

Volume 76, 2017.
K.S. Meyer. Islands in a Sea of Mud: Insights From Terrestrial Island Theory for Community Assembly on Insular Marine Substrata. pp. 1–40.
E.M. Montgomery, J.-F. Hamel, and A. Mercier. Patterns and Drivers of Egg Pigment Intensity and Colour Diversity in the Ocean: A Meta-Analysis of Phylum Echinodermata. pp. 41–104.
F. Espinosa and G.A. Rivera-Ingraham. Biological Conservation of Giant Limpets: The Implications of Large Size. pp. 105–156.
L. Yebra, T. Kobari, A.R. Sastri, F. Gusmão, and S. Hernández-León. Advances in Biochemical Indices of Zooplankton Production. pp. 157–240.

Volume 77, 2017.
Dayv Lowry and Shawn E. Larson. Introduction to Northeast Pacific Shark Biology, Ecology, and Conservation. pp. 1–8.
David A. Ebert, Jennifer S. Bigman, and Julia M. Lawson. Biodiversity, Life History, and Conservation of Northeastern Pacific Chondrichthyans pp. 9–78.
Shawn E. Larson, Toby S. Daly-Engel, and Nicole M. Phillips. Review of Current Conservation Genetic Analyses of Northeast Pacific Sharks. pp. 79–110.
Joseph J. Bizzarro, Aaron B. Carlisle, Wade D. Smith, and Enric Cortés. Diet Composition and Trophic Ecology of Northeast Pacific Ocean Sharks. pp. 111–148.
Jonathan C.P. Reum, Gregory D. Williams, and Chris J. Harvey. Stable Isotope Applications for Understanding Shark Ecology in the Northeast Pacific Ocean. pp. 149–178.
Mary E. Matta, Cindy A. Tribuzio, David A. Ebert, Kenneth J. Goldman, and Christopher M. Gburski. Age and Growth of Elasmobranchs and Applications to Fisheries Management and Conservation in the Northeast Pacific Ocean. pp. 179–220.

Volume 78, 2017.
Shawn E. Larson and Dayv Lowry. Introduction to Northeast Pacific Shark Biology, Research, and Conservation, Part B. pp. 1–8.

Jackie King, Gordon A. McFarlane, Vladlena Gertseva, Jason Gasper, Sean Matson, and Cindy A. Tribuzio. Shark Interactions With Directed and Incidental Fisheries in the Northeast Pacific Ocean: Historic and Current Encounters, and Challenges for Shark Conservation. pp. 9–44.

Dovi Kacev, Timothy J. Sippel, Michael J. Kinney, Sebastián A. Pardo, and Christopher G. Mull. An Introduction to Modelling Abundance and Life History Parameters in Shark Populations. pp. 45–88.

Michael Grassmann, Bryan McNeil, and Jim Wharton. Sharks in Captivity: The Role of Husbandry, Breeding, Education, and Citizen Science in Shark Conservation. pp. 89–120.

Peter A. Mieras, Chris Harvey-Clark, Michael Bear, Gina Hodgin, and Boone Hodgin. The Economy of Shark Conservation in the Northeast Pacific: The Role of Ecotourism and Citizen Science. pp. 121–154.

Dayv Lowry. Conclusions: The Future of Shark Management and Conservation in the Northeast Pacific Ocean. pp. 155–164.

Volume 79, 2018.

Patricia L.M. Lee, Gail Schofield, Rebecca I. Haughey, Antonios D. Mazaris, and Graeme C. Hays. A Review of Patterns of Multiple Paternity Across Sea Turtle Rookeries. pp. 1–32.

Johanne Vad, Georgios Kazanidis, Lea-Anne Henry, Daniel O.B. Jones, Ole S. Tendal, Sabine Christiansen, Theodore B. Henry, and J. Murray Roberts. Potential Impacts of Offshore Oil and Gas Activities on Deep-Sea Sponges and the Habitats They Form. pp. 33–60.

Gianmarco Ingrosso, Marco Abbiati, Fabio Badalamenti, Giorgio Bavestrello, Genuario Belmonte, Rita Cannas, Lisandro Benedetti-Cecchi, Marco Bertolino, Stanislao Bevilacqua, Carlo Nike Bianchi, Marzia Bo, Elisa Boscari, Frine Cardone, Riccardo Cattaneo-Vietti, Alessandro Cau, Carlo Cerrano, Renato Chemello, Giovanni Chimienti, Leonardo Congiu, Giuseppe Corriero, Federica Costantini, Francesco De Leo, Luigia Donnarumma, Annalisa Falace, Simonetta Fraschetti, Adriana Giangrande, Maria Flavia Gravina, Giuseppe Guarnieri, Francesco Mastrototaro, Marco Milazzo, Carla Morri, Luigi Musco, Laura Pezzolesi, Stefano Piraino, Fiorella Prada, Massimo Ponti, Fabio Rindi, Giovanni Fulvio Russo, Roberto Sandulli, Adriana Villamor, Lorenzo Zane, and Ferdinando Boero. Mediterranean Bioconstructions Along the Italian Coast. pp. 61–136.

Alex D. Rogers. The Biology of Seamounts: 25 Years on. pp. 137–224.

Volume 80, 2018.

Michael H. Schleyer and Sean N. Porter. Drivers of Soft and Stony Coral Community Distribution on the High-Latitude Coral Reefs of South Africa. pp. 1–56.

Sara González-Delgado and José Carlos Hernández. The Importance of Natural Acidified Systems in the Study of Ocean Acidification: What Have We Learned? pp. 57–100.

Volume 81, 2018.

Melissa C. Cook, Adam May, Lucas Kohl, Geert Van Biesen, Christopher C. Parrish, and Penny L. Morrill. The Potential Impact of Hydrocarbons on Mussels in Port au Port Bay, Newfoundland pp. 1–22.

Yunwen Tao, Bing Chen, Baiyu (Helen) Zhang, Zhiwen (Joy) Zhu, and Qinhong Cai. Occurrence, Impact, Analysis and Treatment of Metformin and Guanylurea in Coastal Aquatic Environments of Canada, USA and Europe pp. 23–58.

Gong Zhang, Chun Yang, Mariam Serhan, Graeme Koivu, Zeyu Yang, Bruce Hollebone, Patrick Lambert, and Carl E. Brown. Characterization of Nitrogen-Containing Polycyclic Aromatic Heterocycles in Crude Oils and Refined Petroleum Products. pp. 59–96.

Arpana Rani Datta, Qiao Kang, Bing Chen, and Xudong Ye. Fate and Transport Modelling of Emerging Pollutants from Watersheds to Oceans: A Review. pp. 97–128.

Meijin Du, Wenwen Gu, Xixi Li, Fuqiang Fan, and Yu Li. Modification of Hexachlorobenzene to Molecules with Lower Long-Range Transport Potentials Using 3D-QSAR Models with a Full Factor Experimental Design pp. 129–166.

Weiyun Lin, Xixi Li, Min Yang, Kenneth Lee, Bing Chen, and Baiyu (Helen) Zhang. Brominated Flame Retardants, Microplastics, and Biocides in the Marine Environment: Recent Updates of Occurrence, Analysis, and Impacts. pp. 167–212.

Xixi Li, Zhenhua Chu, Jiawen Yang, Minghao Li, Meijin Du, Xiaohui Zhao, Zhiwen (Joy) Zhu, and Yu Li. Synthetic Musks: A Class of Commercial Fragrance Additives in Personal Care Products (PCPs) Causing Concern as Emerging Contaminants. pp. 213–280.

Volume 82, 2019.

Junaid S. Khan, Jennifer F. Provencher, Mark R. Forbes, Mark L. Mallory, Camille Lebarbenchon, and Karen D. McCoy. Parasites of seabirds: A survey of effects and ecological implications. pp. 1–50.

Charlotte Havermans, Holger Auel, Wilhelm Hagen, Christoph Held, Natalie S. Ensor, and Geraint A. Tarling. Predatory zooplankton on the move: *Themisto* amphipods in high-latitude marine pelagic food webs. pp. 51–92.

Angel Borja, Guillem Chust, and Iñigo Muxika. Forever young: The successful story of a marine biotic index. pp. 93–128.

Ferdinando Boero, Francesco De Leo, Simonetta Fraschetti, and Gianmarco Ingrosso. The Cells of Ecosystem Functioning: Towards a holistic vision of marine space. pp. 129–154.

Volume 83, 2019.

Dayv Lowry and Shawn Larson. Introduction: The sharks of Pacific Mexico and their conservation: Why should we care?. pp. 1–10.

Luz Erandi Saldaña-Ruiz, Emiliano García-Rodríguez, Juan Carlos Pérez-Jiménez, Javier Tovar-Ávila, and Emmanuel Rivera-Téllez. Biodiversity and conservation of sharks in Pacific Mexico. pp. 11–60.

Felipe Galván-Magaña, José Leonardo Castillo-Geniz, Mauricio Hoyos-Padilla, James Ketchum, A. Peter Klimley, Sergio Ramírez-Amaro, Yassir Eden Torres-Rojas, and Javier Tovar-Avila. Shark ecology, the role of the apex predator and current conservation status. pp. 61–114.

Jonathan Sandoval-Castillo. Conservation genetics of elasmobranchs of the Mexican Pacific Coast, trends and perspectives. pp. 115–158.

Volume 84, 2019.

Brian Morton and Fabrizio Marcondes Machado. Predatory marine bivalves: A review. pp. 1–98.

Volume 85, 2020.

James T. Ketchum, Mauricio Hoyos-Padilla, Alejandro Aldana-Moreno, Kathryn Ayres, Felipe Galván-Magaña, Alex Hearn, Frida Lara-Lizardi, Gador Muntaner-López, Miquel Grau, Abel Trejo-Ramírez, Darren A. Whitehead, and A. Peter Klimley. Shark movement patterns in the Mexican Pacific: A conservation and management perspective. pp. 1–38.

Oscar Sosa-Nishizaki, Emiliano García-Rodríguez, Christian D. Morales-Portillo, Juan C. Pérez-Jiménez, M. del Carmen Rodríguez-Medrano, Joseph J. Bizzarro, and José Leonardo Castillo-Géniz. Fisheries interactions and the challenges for target and nontargeted take on shark conservation in the Mexican Pacific. pp. 39–70.

A.M. Cisneros-Montemayor, E.E. Becerril-García, O. Berdeja-Zavala, and A. Ayala-Bocos. Shark ecotourism in Mexico: Scientific research, conservation, and contribution to a Blue Economy. pp. 71–92.
Oscar Sosa-Nishizaki, Felipe Galván-Magaña, Shawn E. Larson, and Dayv Lowry. Conclusions: Do we eat them or watch them, or both? Challenges for conservation of sharks in Mexico and the NEP. pp. 93–102.

Volume 86, 2020.
David J. Vance and Peter C. Rothlisberg. The biology and ecology of the banana prawns: *Penaeus merguiensis* de Man and *P. indicus* H. Milne Edwards. pp. 1–140.
Jeff A. Eble, Toby S. Daly-Engel, Joseph D. DiBattista, Adam Koziol, and Michelle R. Gaither. Marine environmental DNA: Approaches, applications, and opportunities. pp. 141–170.

Volume 87, 2020.
Bernhard M. Riegl and Peter W. Glynn. Population dynamics of the reef crisis: Consequences of the growing human population. pp. 1–30.
Chiara Pisapia, Peter J. Edmunds, Holly V. Moeller, Bernhard M. Riegl, Mike McWilliam, Christopher D. Wells, and Morgan S. Pratchett. Projected shifts in coral size structure in the Anthropocene. pp. 31–60.
Alex E. Mercado-Molina, Alberto M. Sabat, and Edwin A. Hernández-Delgado. Population dynamics of diseased corals: Effects of a Shut Down Reaction outbreak in Puerto Rican *Acropora cervicornis*. pp. 61–82.
Elizabeth A. Goergen, Kathleen Semon Lunz, and David S. Gilliam. Spatial and temporal differences in *Acropora cervicornis* colony size and health. pp. 83–114.
Rian Prasetia, Zi Wei Lim, Aaron Teo, Tom Shlesinger, Yossi Loya, and Peter A. Todd. Population dynamics and growth rates of free-living mushroom corals (Scleractinia: Fungiidae) in the sediment-stressed reefs of Singapore. pp. 115–140.
Joshua S. Feingold and Brandon Brulé. Population fluctuations of the fungiid coral *Cycloseris curvata*, Galápagos Islands, Ecuador. pp. 141–166.
Stuart A. Sandin, Clinton B. Edwards, Nicole E. Pedersen, Vid Petrovic, Gaia Pavoni, Esmeralda Alcantar, Kendall S. Chancellor, Michael D. Fox, Brenna Stallings, Christopher J. Sullivan, Randi D. Rotjan, Federico Ponchio, and Brian J. Zgliczynski. Considering the rates of growth in two taxa of coral across Pacific islands. pp. 167–192.
Geórgenes Cavalcante, Filipe Vieira, Jonas Mortensen, Radhouane Ben-Hamadou, Pedro Range, Elizabeth A. Goergen, Edmo Campos,

and Bernhard M. Riegl. Biophysical model of coral population connectivity in the Arabian/Persian Gulf. pp. 193–222.
S.A. Matthews, C. Mellin, and Morgan S. Pratchett. Larval connectivity and water quality explain spatial distribution of crown-of-thorns starfish outbreaks across the Great Barrier Reef. pp. 223–258.
S.A. Matthews, K. Shoemaker, Morgan S. Pratchett, and C. Mellin. COTSMod: A spatially explicit metacommunity model of outbreaks of crown-of-thorns starfish and coral recovery. pp. 259–290.
Tim R. McClanahan. Coral community life histories and population dynamics driven by seascape bathymetry and temperature variability. pp. 291–330.
William F. Precht, Richard B. Aronson, Toby A. Gardner, Jennifer A. Gill, Julie P. Hawkins, Edwin A. Hernéndez-Delgado, Walter C. Jaap, Tim R. McClanahan, Melanie D. McField, Thaddeus J.T. Murdoch, Maggy M. Nugues, Callum M. Roberts, Christiane K. Schelten, Andrew R. Watkinson, and Isabelle M. Côté. The timing and causality of ecological shifts on Caribbean reefs. pp. 331–360.
Howard R. Lasker, Lorenzo Bramanti, Georgios Tsounis, and Peter J. Edmunds. The rise of octocoral forests on Caribbean reefs. pp. 361–410.
Sascha C.C. Steiner, Priscilla Martínez, Fernando Rivera, Matthew Johnston, and Bernhard M. Riegl. Octocoral populations and connectivity in continental Ecuador and Galápagos, Eastern Pacific. pp. 411–442.
Peter W. Glynn, Renata Alitto, Joshua Dominguez, Ana B. Christensen, Phillip Gillette, Nicolas Martinez, Bernhard M. Riegl, and Kyle Dettloff. A tropical eastern Pacific invasive brittle star species (Echinodermata: Ophiuroidea) reaches southeastern Florida. pp. 443–472.

Volume 88, 2021.
Pat Hutchings. Potential loss of biodiversity and the critical importance of taxonomy—An Australian perspective. pp. 3–16.
Ferdinando Boero. Mission possible: Holistic approaches can heal marine wounds. pp. 19–38.
Iraide Artetxe-Arrate, Igaratza Fraile, Francis Marsac, Jessica H. Farley, Naiara Rodriguez-Ezpeleta, Campbell R. Davies, Naomi P. Clear, Peter Grewe, and Hilario Murua. A review of the fisheries, life history and stock structure of tropical tuna (skipjack *Katsuwonus pelamis*, yellowfin *Thunnus albacares* and bigeye *Thunnus obesus*) in the Indian Ocean. pp. 39–90.
Yan Ji and Kunshan Gao. Effects of climate change factors on marine macroalgae: A review. pp. 91–136.

Brian Morton and Fabrizio Marcondes Machado. The origins, relationships, evolution and conservation of the weirdest marine bivalves: The watering pot shells. A review. pp. 139–220.

Volume 89, 2021.
Stanislao Bevilacqua, Laura Airoldi, Enric Ballesteros, Lisandro Benedetti-Cecchi, Ferdinando Boero, Fabio Bulleri, Emma Cebrian, Carlo Cerrano, Joachim Claudet, Francesco Colloca, Martina Coppari, Antonio Di Franco, Simonetta Fraschetti, Joaquim Garrabou, Giuseppe Guarnieri, Cristiana Guerranti, Paolo Guidetti, Benjamin S. Halpern, Stelios Katsanevakis, Maria Cristina Mangano, Fiorenza Micheli, Marco Milazzo, Antonio Pusceddu, Monia Renzi, Gil Rilov, Gianluca Sarà, and Antonio Terlizzi.
Mediterranean rocky reefs in the Anthropocene: Present status and future concerns. pp. 1–52.
Johanne Vad, Kelsey Archer Barnhill, Georgios Kazanidis, and J. Murray Roberts.
Human impacts on deep-sea sponge grounds: Applying environmental omics to monitoring. pp. 53–78.
Angel Borja and Michael Elliott.
From an economic crisis to a pandemic crisis: The need for accurate marine monitoring data to take informed management decisions. pp. 79–114.

Volume 90, 2021.
Shawn Larson, Dayv Lowry, Nicholas K. Dulvy, Jim Wharton, Felipe Galván-Magaña, Abraham B. Sianipar, Christopher G. Lowe, and Erin Meyer. Current and future considerations for shark conservation in the Northeast and Eastern Central Pacific Ocean. pp. 1–50.
Ferdinando Boero. The future ocean we want. pp. 51–64.

Volume 91, 2022.
Jean-François Hamel, Igor Eeckhaut, Chantal Conand, Jiamin Sun, Guillaume Caulier, and Annie Mercier. Global knowledge on the commercial sea cucumber *Holothuria scabra*. pp. 1–286.

Volume 92, 2022.
Monica Montefalcone, Alice Oprandi, Annalisa Azzola, Carla Morri, and Carlo Nike Bianchi. Serpulid reefs and their role in aquatic ecosystems: A global review. pp. 1–54.

Matthew R. Nitschke, Sabrina L. Rosset, Clinton A. Oakley, Stephanie G. Gardner, Emma F. Camp, David J. Suggett, and Simon K. Davy. The diversity and ecology of Symbiodiniaceae: A traits-based review. pp. 55–128.
David Rodríguez-Rodríguez and Javier Martínez-Vega. Ecological effectiveness of marine protected areas across the globe in the scientific literature. pp. 129–154.

Opinion

CHAPTER ONE

Marine mammal conservation in the 21st century: A plea for a paradigm shift towards mindful conservation

Fabian Ritter*
M.E.E.R. e.V., Berlin, Germany
*Corresponding author: e-mail address: ritter@m-e-e-r.de

Contents

1. Preface 4
2. Change 6
3. Three ways of marine (mammal) conservation 8
   3.1 A—Traditional conservation 8
   3.2 B—Contemporary conservation 10
   3.3 C—Mindful conservation 12
4. Mindful conservation and the commons 14
5. The personal (but not necessarily private) dimension 15
6. Conclusion 17
Acknowledgements 18
References 18

Abstract

Marine mammals are regarded in high esteem by the general public, and are recognized as flagship species for conservation, while at the same time they suffer from anthropogenic impacts on a global scale, and often in extreme ways. It seems there is a huge discrepancy between how we humans think about our fellow creatures in the sea, and how we behave to impact and/or conserve them. Here, I examine why the purely scientific and thus intellectual approach to marine mammal conservation has had limited success over the past decades. While there are some obvious success stories in cetacean conservation, the situation today is, for many species and populations, more dire than it has ever been. The idea of 'we need to know more'—a credo of the scientific community—often is politically misrepresented to postpone necessary conservation decisions. To adapt our path towards more profound and, importantly, more effective marine conservation, as conservationists we need to go deeper and change the narrative of separation, i.e., the concept of humans being set apart from the rest of nature. Instead, there is a need to create a narrative of

Advances in Marine Biology, Volume 93
ISSN 0065-2881
https://doi.org/10.1016/bs.amb.2022.09.001

connectedness, i.e., the consciousness of humans being an integral part of the planetary system. Rather than telling horror stories about the plight of marine mammals, conservationists also need to trigger positive emotions about them in ourselves. More holistic aspects of conservation need to be incorporated in our future efforts, including the fuller integration of traditional knowledge and indigenous wisdom, recognizing ecosystem functions of marine life and protecting the processes they sustain, respecting 'holiness' of nature while focusing on the animals' individuality, personhood and the cultural identity of distinct communities. Effective marine mammal conservation will be possible only on the basis of a profound change of our own values and a fundamental change of the societal system we are living in.

1. Preface

The 2022 Annual Conference of the European Cetacean Society (ECS) chose the topical theme 'Marine Mammal Conservation: Are we on the right path?'. Having spent a large part of my career dealing with marine conservation focusing on cetaceans, I submitted a talk which wasn't grounded on presenting scientific findings, but on reflecting on my roughly 30 years of work in the field of marine (mammal) research and conservation. The conference organizers accepted the talk and generously allowed it to be presented to the audience at the beginning of the event. This article is based on that talk. I will be quite philosophical, rather than presenting data. One could say I will be speaking (writing) from my heart rather than taking a purely intellectual approach. I have been inspired by the article *Combining Love and Knowledge to heal the Ocean* (Lotze, 2020)—which takes a similar approach to the one that I am elaborating here.

The short answer to the question 'Are we on the right path?' in my eyes is *No*. The longer answer has to do with the fact that we as scientist often do *not* speak from our hearts. My own case is pretty straightforward: it has always been my passion and love for cetaceans that drove me. As a matter of fact, I started dreaming regularly about dolphins when I was young, a recurring theme that finally had me reading books and watching documentaries about these animals. This eventually led into the study of biology and a profession focusing on scientific studies on cetaceans, and on marine mammal conservation. I am fairly sure that this same love and enthusiasm is the driving force for many of my colleagues, too.

The baseline is that we are in a dire situation. Not only many marine mammals, but also us as mankind: we are in deep trouble. A time has come where there is an urgent need to turn the wheel on a global scale. Regarding cetaceans, the outlook of many of the roughly 90 species and numerous populations today seems to be less hopeful than it has ever been. That is true, despite the fact that a number of large whale population have recovered quite well over the past decades (Lotze et al., 2011; Magera et al., 2013). Nevertheless, marine pollution, bycatch in fishing gear, underwater noise, shipping traffic, overfishing, climate change and a whole suit of other threats (Avila et al., 2018; Nelms et al., 2021) plus the cumulative and often synergistic effects (Wright and Kyhn, 2015; Judd et al., 2015) are hampering a healthy development of dolphins, whales and porpoises on a global scale—as well as other marine life and the ecosystems they are living in (e.g. see McCauley et al., 2015; Thomas et al., 2016). On a global scale, we are witnessing an overall decline of health in nature.

Here are two examples of conservation efforts I personally have been involved in over the past decades. The first is about harbour porpoises (*Phocoena phocoena*) living in the waters of the European Union (EU). Notably this is a species strictly protected through the EU Habitats Directive and the EU Marine Strategy Framework Directive, but also many other Agreements such as HELCOM and OSPAR (Carlén et al., 2021). Europe-wide, porpoises are dying in the thousands every year due to anthropogenic impacts, primarily as victims of the most prevalent direct threat: bycatch in fishing gear (Dolman et al., 2016; Rogan et al., 2021). The subpopulation in the Baltic Proper is critically endangered (Hammond et al., 2016), and the status of the Iberian subpopulation appears to be similar (Carlén et al., 2021). Do we see significant progress in conserving this species? I don't think so, despite International Agreements (including a dedicated treaty on the protection of small cetaceans, the *Agreement on the Conservation of Small Cetaceans in the Baltic, North East Atlantic, Irish and North Seas and North Sea and Contiguous Areas*, ASCOBANS), existing national and EU-wide policies, intense research efforts, and a lot of public education by a large number of Non-Governmental Organizations (NGOs).

The second example concerns ship strikes: Since I started to get involved in this issue some 20 years ago, the global number of cetaceans being hit by vessels seems to be ever increasing (Laist et al., 2001; Ritter and Panigada, 2018; Schoeman et al., 2020; van Waerebeek et al., 2007), making it another prevalent threat that these animals have to face today. In the Canary Islands where our research group is conducting long-term cetacean research, the problem arose in the late 1990s, and has increased steadily until today, notwithstanding a number of initiatives to foster mitigation measures and relevant conservation steps (Carrillo and Ritter, 2010; Ritter, 2010).

The overall situation of cetaceans world-wide led to an unprecedented statement of concern from >350 marine mammal experts issued in 2020 shortly after the Second World Marine Mammal Conference (Rochas-Bracho et al., 2020)[a]. The experts noted that the situation for many populations and species is 'now critical' and that 'The lack of concrete action to address threats adversely affecting cetaceans in our increasingly busy, polluted, over-exploited and human-dominated seas and major river systems, means that many, one after another, will likely be declared extinct within our lifetimes'.

So, what's going wrong in marine (mammal) conservation, despite all the conservation efforts and the many commitments from the political level, from industries, from scientists, and from NGOs?

2. Change

Everyone would probably agree that changing our human behaviour is key to initiating a transformation of the path we are currently collectively pursuing. But changing both our individual and our collective behaviours seems to be the crux here. Is changing behaviour based on the amount of knowledge we have? We, as scientists, are trained to believe in the concept that more information (more data) will lead to better solutions—and in a way, that's our job. But the idea of 'We need to know more'—which is a logical stand of the scientific community—today often is even being politically misrepresented in order to postpone overdue conservation decisions. Whether it is the designation of a marine protected area (MPA), the determination of fishing quota or the calculation of bycatch rates of non-target species in fisheries, often it is stated that more research, and thus more knowledge is needed to achieve such goals. It may well turn out that this is a never-ending story. While scientists are not decision makers, they can, however, emphasize the need for swift action even in the absence of absolute knowledge. Missing knowledge should not need to lead to procrastination and in many cases, it just might not be feasible to fully determine scientifically what is being looked for. Hence, '*generating a new world view of so much respect for nature that we radically alter our present ways of subjugating and diminishing it in the name of supposed human progress*' (Würsig, 2020, p. 25) becomes a necessity.

[a] See https://static1.squarespace.com/static/5b8797f45417fc360d47af9f/t/607dce210a2ec22f63a04da0/1618857508201/Scientists+Sign+on+Final_MasterDocument-uPDATED.pdf.

It must be assumed that there is no 'automatic' relationship between understanding and (re)action. In my opinion, the idea of 'We need to know more before we can act' has often failed over the past decades, when it comes to marine conservation (compare Bearzi, 2020). While scientists are not the ones taking conservation decisions, they must be aware that decision makers may use the apparent need for more knowledge to delay efforts to protect nature or implementing scientific results.

The main reason for the demise of our planet is that it is rooted in an even more profound and dominating story of our time: the narrative of Separation. Setting us humans apart from nature, considering nature to be just a place of resources that can and must be exploited, is a prevailing human attitude, at least of those in the modern, 'civilized' world. Our lives are organized around the pursuit of wealth (or the safeguarding of acquired wealth) which is achieved by direct or indirect consumption, i.e., exploitation of 'natural resources', as we call it. Often, nature is even perceived as an enemy—e.g., wilderness for many has a negative connotation, and being in nature without any safety-lines connecting us to civilization will cause unease and fear in many of us. The COVID pandemic is a good example of how humankind is dealing with a perceived threat that slips out of control. Soon after the start of the pandemic, state leaders *declared war* against the virus and an accompanying bellicose language has been built around our strategies to deal with the pandemic (Panzeri et al., 2021) ever since, often depicting a fight where the 'enemy' (the virus) has to be destroyed and overcome.

What is more, our modern societal system is largely based on the longing for more, better, bigger. As individuals and societies, we want to reach goals, become more effective, more wealthy, more beautiful, more happy, more advanced, etc. The myth of ever-lasting growth is being pursued as the overall dictum of our times, despite the fact that the fundamental error in this narrative was disclosed already half a century ago by the famous Club of Rome (https://www.clubofrome.org/publication/the-limits-to-growth/). Tragically, setting ourselves apart from nature also led us to separating ourselves from each other—as communities, nations, races, genders, scientific disciplines, you name it. That path must come to an end. But how do we get there?

All the big crises we are facing today—including climate change and the subsequent sea level rise, marine pollution, biodiversity loss, even the COVID pandemic, and yes, also wars—have something in common: They exemplify our societies' setup which detaches ourselves from the natural and ecological processes on this planet. In my eyes, all these crises are not *causes* for the situation we are in, they are *symptoms* of an even bigger crisis. A crisis of our consciousness.

The famous Vietnamese Buddhist monk Tich Nhat Hanh (1926–2022) was an advocate of mindfulness, and he suggested that a mindful attitude will contribute to creating a better world. Mindfulness is an attitude focusing on the present moment and it is based on a good balance between mind and heart. It encompasses all facets of our perception, including the intellect and our thinking in general, our emotions and intuitions, as well as our beliefs. Thus, mindfulness touches on us in a truly holistic way. When looking for solutions, it doesn't just scratch the surface, but aims deeper at the root causes of our behaviours by uncovering the underlying psychological processes. Thus, it identifies the ideas, beliefs, and the story our actions are based on. Changing such beliefs is the precondition of long-lasting change. I believe that we can learn a lot from that philosophy. So, how can we incorporate mindfulness into marine conservation?

3. Three ways of marine (mammal) conservation

Broadly speaking, we can distinguish three ways of marine conservation, and they all have an underlying narrative:

- **A**—Traditional conservation, following the narrative of separation ('humans against nature')
- **B**—Contemporary conservation, based on a rather utilitarian narrative ('nature at service of humans'), while
- **C**—Mindful conservation, embedded in the narrative of connectedness ('humans as integral part of nature')

Obviously, these categories are not strictly separate from each other, rather they are connected, there is overlap, and they can be seen as building upon each other. I postulate that we currently are somewhere between A and B, and that we have to get to C to really turn the tide in the marine conservation realm.

3.1 A—Traditional conservation

Table 1 gives an overview of the three ways of conservation based on examples from my field of work, cetacean conservation. Traditionally, through a mechanistic and data-centric worldview, conservation was (and often still is) pursued in a purely rational way and grounded on the assumption that nature can be looked at objectively. The method that is assumed to be adequate to

Table 1 The three types of conservation as laid out in this paper: (A) Traditional, (B) Contemporary, (C) Mindful.

	A—Traditional	**B—Contemporary**	**C—Mindful**
Worldview	Mechanistic, reductionist, rational,	Eco-logical	Holistic
	Data-focused	Utilitarian	Traditional knowledge
	More knowledge = solution	Nature at service	Ecological wisdom
Narrative	*Separation*	*Interrelation*	*Connectedness*
	Humans against nature	*Nature at human's service*	*Humans as part of nature*
Units to conserve	Species/population/ (Sub-population)	Species/populations and their habitats as ecological entities (e.g. island waters, seamounts, canyons, etc.)	Ecosystems, plus animals cultural communities (families, matrilines) and individuals (intrinsic value, personality)
Rationale (*What matters?*)	Population size	Ecosystem functions	Connectedness
		Ecosystem services	Relation to 'big picture' Planetary health
Level of action	Protection by space and numbers	Precaution, ecosystem-based approach ecosystem services (incl. carbon values)	(Overarching) legal frameworks to conserve ecological processes, holiness and wilderness
	Management of human behaviour as 'threat control'		Rights for nature

understand natural processes is cutting them down into pieces and trying to understand those so as to explain the phenomena we observe (reductionism). The focus predominantly has been on species and populations, with population *size* often as the key determining variable, complemented by other quantitative approaches. Accordingly, managing human behaviour has been based on reaching numerical values, such as numbers of animals within a species or population, or thresholds for 'acceptable' levels of anthropogenic impacts. 'Data is everything'—is the reality of conservation policies in most countries until today. The Potential Biological Removal (PBR) approach enshrined in the US Marine Mammal Protection Act is a good example. The PBR aims to determine the number of animals that can be 'safely' removed from a given marine mammal population. 'Removal' thereby is defined as any death caused by anthropogenic impacts, such as an animal fatally entangled in a net or hit by a vessel. The underlying logic is that a population can withstand the impacts as long as the PBR is not exceeded. Another example is the Catch Limit Algorithm (CLA) developed under the auspices of the International Whaling Commission (IWC). Similar to PBR, the CLA calculates a threshold that, if not exceeded, will make whaling operations sustainable. In these contexts, natural whale populations often are referred to as 'stocks' which further illustrates the thinking behind such frameworks. In the EU, there is currently also a strong trend to establish 'sustainable removal rates' for marine mammal population similar to PBR and CLA, a development much criticized by conservationists because it leaves out any consideration of animal welfare (compare Clegg et al., 2021; Papastavro et al., 2017). Overall, the traditional strategy of conserving marine 'resources' follows the idea of reducing human impacts, a 'threat control approach', as it were, rather than preventing them in the first place. The big picture thus often has been missing in traditional conservation, especially at the management and implementation level.

3.2 B—Contemporary conservation

Today, ecosystem-based management (EBM) is slowly gaining ground within the international marine management and conservation arena, because it became obvious that a species or population has to be protected together with its habitats, simply because species and ecosystems cannot be separated from each other. Hence, the ecological relationships increasingly come to the fore, and the intertwined nature of habitats as well as their species diversity are to be conserved. The goal is to determine representative fractions of ecosystems so as to protect them, e.g., by delineating a critical habitat as an MPA. Still rather data-centric, such contemporary approaches focus more on ecological entities. Furthermore, we begin to comprehend that cetaceans, and other marine life, are critical elements of marine ecosystems. They fulfill a whole range of ecosystem functions and "deliver" many "ecosystem services" to us (Roman and McCarthy, 2010; Roman et al., 2014; IWC, 2021)—it seems that they even can help us in combating climate change (Chami et al., 2019; Pershing et al., 2010). In that sense, it becomes clear that separating species from their natural environment has not been the ideal concept.

It is now acknowledged that safeguarding ecosystems at a larger scale will also benefit ourselves. As an example, it has been calculated that a single whale can have a monetary value of about 2 million US$, based on the ecosystem service it delivers and its significance for the whale watching industry (Chami et al., 2019). Additionally, whales are accumulating large amounts of carbon in their bodies, and much of that carbon will be 'sequestered' into the deep sea when the whale dies and its carcass sinks to the sea floor (IWC, 2021). Such whale falls additionally create very diverse micro-habitats in great depths and might also be serving as stepping stones for the distribution of various deep-sea species (Smith et al., 2015). Shedding light on the services of natural ecosystems (and larger parts thereof such as whale populations) has become very common in the climate crisis debate, and so-called Nature Based Solutions (NBS)—i.e., regarding mangrove forests, seagrass meadows or populations of large marine vertebrates as effective carbon sinks—are at the core of such discussions. *Blue Carbon* has become a new buzz word.

While all this must be welcomed, such approaches that put a 'price tag' on nature, a concept which is now gaining a lot of traction, and evaluating its "services" in a rather economic way clearly follows a utilitarian narrative. There is a danger of focusing too strongly on monetary values, thereby disregarding the fact that nature can be valued in many other ways. We are doing nature no justice when we consider it only a variable in our economic thinking and a tool in the tool-box of solutions for the climate crisis. Nature is more than just 'at our service', natural places and beings have an *intrinsic value* irrespective of their usefulness for us humans. Cetaceans not only have intelligence and culture, but also *dignity* (White, 2021). Nature must not be saved for the sake of humans, but for its own sake.

Another significant facet of contemporary conservation is the precautionary principle. Importantly, the precautionary approach steps away from the philosophy of 'We need more data'. Explicitly, it needs to appear on the stage of conservation when there is a clear need for protecting species and habitats even in the absence of full scientific proof and solid data. However, my own experience during negotiations with industry representatives, federal agencies and within international fora such as the IWC and ASCOBANS led me to the conclusion that precaution today often is not being applied. Rather, it seems to be something willingly referred to, but often remaining just something being paid lip service to.

3.3 C—Mindful conservation

Mindful conservation, to the contrary, goes beyond all of this. The overarching guiding principle of mindful conservation is planetary health, i.e., *One Health*—a concept now being increasingly applied on an international level (e.g. see https://www.who.int/news-room/questions-and-answers/item/one-health). Fig. 1 illustrates the relationship between the three ways of conservation, showing that the more mindful conservation efforts are, the more inclusive they become, because they appreciate the complexity of life and its processes, rather than simplifying in a reductionist way.

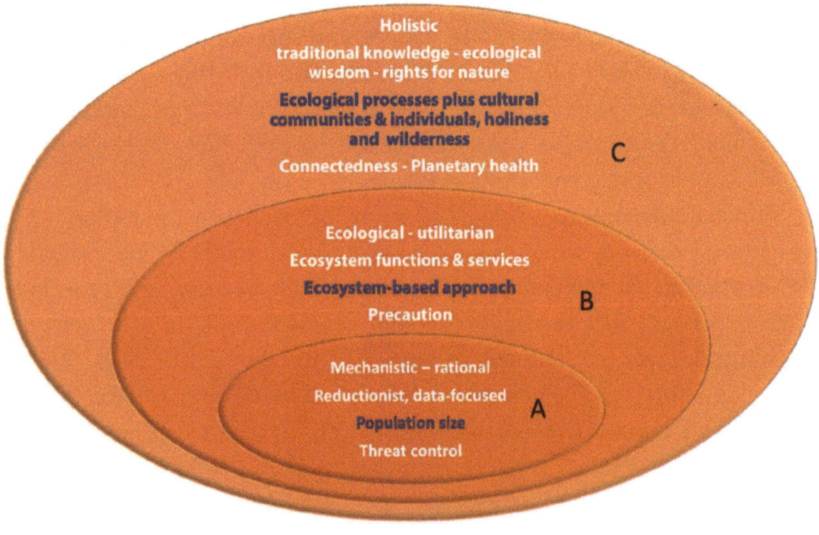

Fig. 1 The three types of conservation as laid out in this chapter: (A) Traditional, (B) Contemporary, (C) Mindful. The graph highlights that the more mindful conservation efforts (C) are, the more they encompass aspects of types (B) abd (A). Blue font = level of action (compare Table 1).

More concretely, and with reference to marine mammals, the following aspects, in my opinion, need to be incorporated in future efforts for conservation to become (more) mindful:

- *Integration of traditional knowledge and indigenous wisdom*, on top of scientific data and the ecosystem-based approach. It is increasingly being recognized that such knowledge provides a unique and rich source of information on biodiversity, behaviour and the biology of cetaceans which can be used to foster conservation efforts. (Here, I am not referring to traditions to herd and hunt cetaceans to kill them, such as in the Faroe Islands and Taiji/Japan or the various kinds of whaling traditions that still exist).

  Stakeholder participation, an important aspect when it comes to conservation efforts—such as establishing MPAs, fisheries regulation or developing guidelines for sustainable tourism—should encompass local people's traditional knowledge and indigenous peoples' wisdom while acknowledging their connection to the land- and seascapes, e.g., by way of surveys, interviews, or co-authorship.
- *Recognition of ecosystem functions of marine life and protecting the processes they sustain,* without a focus on their economic value. It is important to give emphasis on the interrelations between species and those between species and habitats. For example, these can be expressed as the nutrient flux between trophic levels or the carbon cycle they sustain. Looking at complex ecological networks doesn't make things easier, and that is the reason why so many scientists prefer the reductionist way—it makes our professional life easier. The full complexity of nature should not hold us off trying to integrate it into our scientific thinking, though.
- *Appraisal of the uniqueness and 'holiness' of natural places and beings* to further underpin management measures. Many indigenous people build direct relationships to natural features such as landscapes and species, cherishing their spiritual relationship and ecological interdependence. Such appreciation should be integrated into conservation efforts, e.g., by protecting locations and areas based on cultural and spiritual ethics. Assigning a protection status to places of special (cultural and/or spiritual) significance, would be an example of relying on other tenets rather than a wealth of data.
- *Respecting the animals' individuality and personhood.* It is increasingly being acknowledged that animals within a population differ individually, and fulfill specialized roles in their respective social networks (CMS, 2018). As a conceivable consequence, assigning rights to cetaceans, such as the

right to live undisturbed and unharmed, would demonstrate our respect in a new and more profound way. While this may have been seen as far-fetched until recently, there are now many examples world-wide which can be drawn from (see White, 2021), showing that the legal frameworks are shifting into new directions.

- *Letting nature be, i.e., letting wilderness rule, and pursuing a hands-off approach* while relying on the self-healing potential of the seas, will be crucial in many areas. This can be achieved through the establishment of 'no-take zones' which to date are extremely rare even within MPAs or otherwise protected areas (see https://mpatlas.org/zones/). The call from scientists for marine areas where humans are kept out completely must become stronger, and it must be better heard at the management level.
- *Treasuring the cultural identity of distinct animal communities.* Our knowledge about animal cultures is growing steadily. Such new insights must be incorporated in conservation endeavours by targeting cultural entities within larger populations (such as the different orca, *Orcinus orca,* ecotypes. Studying cultural traits of animals can become an important scientific discipline and cetaceans are the perfect fit for such studies.

Considering the latest findings around cetaceans' cognitive capabilities (e.g. Bearzi and Stanford, 2010; Marino, 2002; Marino et al., 2007; White, 2021), sentience (Bearzi et al., 2018; White, 2021), social complexity (Brakes et al., 2021; King et al., 2018) and culture(s) (Whitehead and Rendell, 2015; Fox et al., 2017; Brakes et al., 2019, 2021), I propose that we should look at whales and dolphins as the *indigenous people of the seas*—and treat them accordingly. As soon as we are following such lines of thought, our whole relationship with them can then become a different story.

4. Mindful conservation and the commons

Thich Nhat Hanh has used the term *Interbeing* to describe the connection between living beings, which represents a deeply ecological understanding (in fact, Thich Nhat Hanh wrote a number of books about the global ecological crisis and how to deal with it, e.g. Nhat Hanh, 2021). The current state of our modern world and the disastrous consequences of our focus on competition and individuality would benefit from a move towards more 'Interbeing-ness'. After all, we are realizing that humans are just another part in the intricate networks of ecosystems and all forms of life (Lent, 2021).

Both humans, through their conscious and conscientious decision making, and cetaceans (or more broadly speaking, marine megafauna), through their ecosystem functions and services, can contribute a lot to the well-being of the oceans. In that sense, they all contribute to what now is termed *One Health.* Although this general concept is focused in the well-being of humans and thus follows a anthropocentric path, regarding the planet as a common good that all living beings are sharing does imply that conserving the seas, and building up *One Health* is a contribution from many sides, including animals, plants, fungi and the natural network/ecosystems they create and sustain. Caring for the Commons then becomes a matter of caring for other beings, and in doing so letting other beings help ourselves to heal and restore nature where that has become necessary. Following this line of thought, I believe it makes sense to define 'the Commons' in a new way by expanding the concept into the non-human realm. Broadening its scope from considering only the utility of common goods by humans and shifting towards a stronger focus on their value for the well-being of nature at large would also embed our role as humans into a larger, ecologically based philosophy. Therefore, I suggest the following expanded definition of the Commons:

All goods originating from the earth and serving the life on it, as well as all goods originating from life maintaining itself, shall be treated as common goods serving planetary health.

From such an ecological and holistic understanding arises the need—and the mandate—to care for all living beings appropriate to the specific ecological needs of each of its kind. In this way, our human role can shift from *ruler* or *dominator* to *caretaker* or *guardian* of the planet. This embodies the consideration that humans are inseparable from nature and that our behaviour is complexly connected to all other living beings.

5. The personal (but not necessarily private) dimension

Mindfulness obviously doesn't fall from the sky. On the individual level, it is the result of continuous mental and physical practice. But how can we as scientists or conservationists practice mindfulness and strive to integrate it on a societal level, i.e., into policy, in a pragmatic manner?

I am convinced that making the first steps on a personal level will be key. Change always starts with oneself, and may spread from just a few to many by setting an example. Our search for truth and meaning, as scientists, always is embedded into our individual way to look at the world. Mindfulness, as a practice, likewise has a concrete relationship to our personal development. It is related to individual exercise of e.g., meditation and breathwork. While I do not expect the readers of this article to start sitting in silence and focusing on their breath (although, from my personal experience, I can recommend it), the core quality to be cultivated here is to explore one's own place within the greater context. 'Everything is bound to everything' is a basic truth and while it has almost become a common place, it is also one of the basic laws of ecology as a scientific discipline. I find this an interesting homology between science and spiritual traditions, very much resonating with the idea of mindfulness.

We, as scientists and conservation biologists, chose to deal with the ecological threats to our study subjects in a rational way by collecting data, writing articles, recommending conservation measures to decision makers or engaging in some form of public education. Yet, there is another way to get our bearings within a global crisis: to feel it. As scientists, we are trained not to refer to our emotions and our often-intimate relationship to the 'objects' we study. I believe that the time is ripe to overcome this frontier. Marine conservationists by definition care a lot for the seas and nobody should hold them from expressing what they feel, and talk about it also in the scientific arena which otherwise runs the risk to be seen as 'sterile' and bare of humaneness (compare Bearzi, 2020). Scientific study and emotions are intricately connected, not separated.

Confronting our own emotions in the face of bad things happening such as a dolphin species going extinct, a marine habitat being destroyed or a coastline disappearing due to sea level rise is not an easy task, and psychologically there are good reason to hide away from them. We may experience different phases of dealing with bad news, such as shock, neglection, anger, or grief, etc. Such emotions often are not easy to withstand, but at the same time these very emotions can become a very valid and powerful guide to take action, and to convince others to take action.

Accordingly, I suggest a change of perspective to create a deeper personal connection to nature in the face of the ecological crisis: Any wound that humans have caused to nature can be looked at as *our own* wounds. This new angle can instantly lead to a greater connectedness and in turn creates a deeper motivation in ourselves to heal—the wounds within as well as the wounds we have left on earth. Or, as Jeremy Lent has put it: 'If you believe that all living beings are family, you will treat them in a different way than if you think that the natural world is a resource to be exploited' (Lent, 2021, p.5). While accepting and actually enduring the reality of our emotions that are related to the wounds we experience, we can build the basis of an authentic personal (but not just private) relationship between ourselves and the world around us. This is how I would interpret Thich Nhat Hanh's *Interbeing*.

6. Conclusion

What I am advocating here is a 21st century paradigm shift in marine (mammal) conservation: A transformation of our collective narrative from *Separation* to *Connectedness*. However, it remains to be seen if our societies (and science, for that matter) can make that shift.

Academically, there is always the chance to refer to knowledge that is not purely rational and scientific, such as local ecological knowledge. There are now many approaches visible which I perceive as signs of hope. For example, it is good to see that the integration of traditional knowledge and indigenous wisdom is currently making its way into scientific thinking, and into environmental scientific literature (Carter and Nielsen, 2011; Noongwook et al., 2007; Singleton et al., 2021; Van der Wal et al., 2022). To build bridges between scientific knowledge and indigenous wisdom is exactly what I am advocating here: finding the connectedness where previously we have seen a gap.

As another example of scientific innovation, the International Whaling Commission has released a Resolution on the issue of ecosystem functioning of cetaceans (IWC, 2016), and in 2021, together with the Convention on the Protection of Migrating Species (CMS) hosted the first international expert workshop on this issue (IWC, 2021). Furthermore, as an important step, CMS has integrated into the framework of its work the idea that cultural entities in the cetaceans need to be recognized (CMS, 2014, 2017, 2018), and in the future must be protected in accordance, with attention given to the specialized role individuals play in their societies (CMS, 2018). Notably, here scientists are moving away from regarding cetaceans as identical specimens within a population, rather recognizing the individuals' biographies and their specific social roles. It is our choice (and our responsibility) to integrate these new ideas into our work or not. Thinking outside the box is becoming more and more popular, advanced and acknowledged. We should be courageous enough to express the full range of our ideas, even if they sometimes may be utopic, unrealistic or somewhat 'irrational'.

Science alone is not enough to save the oceans and neither is policy—wisdom and vision need to be combined with the two to initiate the necessary changes. As conservationists, rather than telling never-ending horror stories about the plight of marine mammals, we also need to trigger *positive* emotions in humans. In other words: we need to stop fighting and focus more on our connection with marine mammals, cherishing the astounding similarities between us and them. Emphasis needs to be given to the many benefits that can be gained from protecting marine wildlife and the seas at large, including their value in delivering food for billions of people, as recreational space, their importance for emotional and spiritual well-being and inspiration, the oceans' significance as the largest climate driver on earth, and so forth.

In this way we can heal the oceans by healing our relationship with them. To quote Heike Lotze from her 2020 article: '*Combining our emotional and rational sides can inspire our thinking and transform our actions in new, creative ways. In order to affect wide-ranging and long-lasting change, love and knowledge need to work together to transform our collective relationship with the ocean.*'

To conclude, in my eyes, more effective marine conservation will be possible only on the basis of a profound change of our values and beliefs and a fundamental change of the societal system we are living in, and the narrative it is based on. Such a shift will not come overnight, but rather we must start now to see such changes over the coming years, decades and generations. Signs are that the scientific community is moving in that direction. I am hopeful this article is a contribution to that development.

Acknowledgements

I am very grateful to Charles Sheppard who encouraged me to write this article and publish it, similarly to many colleagues who expressed their appreciation after my talk at the 2022 ECS Annual Conference. Many thanks to Philippa Brakes, Lori Marino, Heike Lotze, Thomas I. White, Mark Simmonds and Giovanni Bearzi who provided very useful comments and suggestions to an earlier manuscript.

References

Avila, I.C., Kaschner, K., Dormann, C.F., 2018. Current global risks to marine mammals: taking stock of the threats. Biol. Conserv. 221, 54–58. https://doi.org/10.1016/j.biocon.2018.02.021.

Bearzi, G., 2020. Marine biology on a violated planet: from science to conscience. Ethics Sci. Environ. Polit. 20, 1–13. https://doi.org/10.3354/esep00189.

Bearzi, M., Stanford, C., 2010. A bigger, better brain. Observations of chimpanzees and dolphins strengthen the notion that humanlike intelligence may not be uniquely human. Am. Sci. 98, 402–409.

Bearzi, G., Kerem, D., Reeves, R.R., 2018. Whale and dolphin behavioural responses to dead conspecifics. Zoology 128, 1–15.
Brakes, P., et al., 2019. Animal cultures matter for conservation. Science. https://doi.org/10.1126/science.aaw3557.
Brakes, P., et al., 2021. A deepening understanding of animal culture suggests lessons for conservation. Proc. R. Soc. B 288, 20202718. https://doi.org/10.1098/rspb.2020.2718.
Carlén, I., Nunny, L., Simmonds, M.P., 2021. Out of sight, out of mind: how conservation is failing European porpoises. Front. Mar. Sci. 8, 617478. https://doi.org/10.3389/fmars.2021.617478.
Carrillo, M., Ritter, F., 2010. Increasing numbers of ship strikes in the Canary Islands: proposals for immediate action to reduce risk of vessel-whale collisions. J. Cetacean Res. Manage. 11, 131–138.
Carter, B.T.G., Nielsen, E.A., 2011. Exploring ecological changes in Cook Inlet beluga whale habitat though traditional and local ecological knowledge of contributing factors for population decline. Marine Policy 35 (2011), 299–308.
Chami, R., Cosimano, T., Fullenkamp, C., Oztosun, S., 2019. Nature's solution to climate change. A strategy to protect whales can limit greenhouse gases and global warming. Int. Monet. Fund Fin. Dev. 56 (4), 34–38.
Clegg, I.L.K., Boys, R.M., Stockin, K.A., 2021. Increasing the awareness of animal welfare science in marine mammal conservation: addressing language, translation and reception issues. Animals 11, 1596. https://doi.org/10.3390/ani11061596.
CMS, 2014. Report of the CMS Scientific Council Workshop on the Conservation Implications of Cetacean Culture. 18 CMS Document UNEP/CMS/Cop11/Inf. 27 pp.
CMS, 2017. UNEP/CMS/Resolution 11.23 (Rev.COP12): Conservation Implications of Animal Culture and Social Complexity.
CMS, 2018. Report of the CMS Workshop on Conservation Implications of Animal Culture and Social Complexity. CMS Document UNEP/CMS/ScC-SC3/Inf.8. 41 pp.
Dolman, S., Baulch, S., Evans, P.G.H., Read, F., Ritter, F., 2016. Towards an EU action plan on cetacean bycatch. Marine Policy 72, 67–75. https://doi.org/10.1016/j.marpol.2016.06.020.
Fox, K.C.R., Muthukrishna, M., Shultz, S., 2017. The social and cultural roots of whale and dolphin brains. Nat. Ecol. Evol. 1, 1699–1705. https://doi.org/10.1038/s41559-017-0336-y.
Hammond, P.S., Bearzi, G., Bjørge, A., Forney, K.A., Karczmarski, L., Kasuya, T., Perrin, W.F., Scott, M.D., Wang, J.Y., Wells, R.S., Wilson, B., 2016. *Phocoena phocoena* (Baltic Sea subpopulation) (no. e. T17031A98831650). The IUCN Red List of Threatened Species.
IWC, 2016. Resolution on Cetaceans and Their Contribution to Ecosystem Functioning.
IWC, 2021. Report of the IWC-CMS Workshop on Cetacean Ecosystem Functioning, Virtual 9-21 April 2021. IWC Document SC/68C/REP/03. 39 pp.
Judd, A.D., Backhaus, T., Goodsir, F., 2015. An effective set of principles for practical implementation of marine cumulative effects assessment. Environ. Sci. Policy 54, 254–262.
King, S.L., Friedman, W.R., Allen, S.J., Gerber, L., Jensen, F.H., Wittwer, S., Connor, R.C., Krützen, M., 2018. Bottlenose dolphins retain individual vocal labels in multi-level alliances. Curr. Biol. https://doi.org/10.1016/j.cub.2018.05.013.
Laist, D.W., Knowlton, A.R., Mead, J.G., Collet, A.S., Podesta, M., 2001. Collisions between ships and whales. Mar. Mamm. Sci. 17, 35–75. https://doi.org/10.1111/j.1748-7692.2001.tb00980.x.
Lent, J., 2021. The Web of Meaning. Profile Books, London. ISBN 978-1-7-565-5.
Lotze, H.K., 2020. Combining love and knowledge to heal the ocean. Ethics Sci. Environ. Polit. 20, 33–39. https://doi.org/10.3354/esep00193.

Lotze, H.K., Coll, M., Magera, A.M., Ward-Peigel, C., Airoldi, L., 2011. Recovery of marine animal populations and ecosystems. Trends Ecol. Evol. 26. https://doi.org/10.1016/j.tree.2011.07.008.

Magera, A.M., Mills Flemming, J.E., Kaschner, K., Christensen, L.B., Lotze, H.K., 2013. Recovery trends in marine mammal populations. PLoS One 8 (10), e77908. https://doi.org/10.1371/journal.pone.0077908.

Marino, L.A., 2002. Convergence of complex cognitive abilities in cetaceans and primates. Brain, Behav. Evol. 59, 21–32.

Marino, L., Connor, R.C., Fordyce, R.E., Herman, L.M., Hof, P.R., et al., 2007. Cetaceans have complex brains for complex cognition. PLoS Biol. 5 (6), e139. https://doi.org/10.1371/journal.pbio.0050139.

McCauley, D.J., et al., 2015. Marine defaunation: animal loss in the global ocean. Science 347, 1255641. https://doi.org/10.1126/science.1255641.

Nelms, S.E., et al., 2021. Marine mammal conservation: over the horizon. Endangered Species Res. 44, 291–325. https://doi.org/10.3354/esr01115.

Nhat Hanh, T., 2021. Zen and the Art of Saving the Planet. Harper Collins. ISBN 978-0-06-295479-4.

Noongwook, G., The Native Village of Savoonga, The Native Village of Gambell, Huntington, H.P., George, J.C., 2007. Traditional knowledge of the bowhead whale (*Balaena mysticetus*) around St. Lawrence Island, Alaska. Arctic 60 (1), 47–54.

Panzeri, F., Di Paola, S., Domaneschi, F., 2021. Does the COVID-19 war metaphor influence reasoning? PLoS One 16 (4), e0250651. https://doi.org/10.1371/journal.pone.0250651.

Papastavro, V., Leaper, R., Lavigne, D., 2017. Why management decisions involving marine mammals should include animal welfare. Marine Policy 79, 19–24.

Pershing, A.J., Christensen, L.B., Record, N.R., Sherwood, G.D., Stetson, P.B., 2010. The impact of whaling on the ocean carbon cycle: why bigger was better. PLoS One 5 (8), e12444. https://doi.org/10.1371/journal.pone.0012444.

Ritter, F., 2010. A quantification of ferry traffic in the Canary Islands (Spain) and its implications for collisions with cetaceans. J. Cetacean Res. Manage. 11 (2), 139–146.

Ritter, F., Panigada, S., 2018. Collisions of vessels with cetaceans—The underestimated threat. In: Shepperd, C. (Ed.), World Seas–An Environmental Evaluation–Volume III: Ecological Issues and Environmental Impacts. Elsevier Academic Press, pp. 531–547.

Rochas-Bracho et al., 2020. The Real And Imminent Extinction Risk To Whales, Dolphins And Porpoises: An Open Letter From Cetacean Scientist. This letter was first compiled during the 2nd World Marine Mammal Conference, Barcelona, December 9-12, 2019.

Rogan, E., Read, A.J., Berggren, P., 2021. Empty promises: the European Union is failing to protect dolphins and porpoises from fisheries by-catch. Fish Fisheries 22, 865–869.

Roman, J., McCarthy, J.J., 2010. The whale pump: marine mammals enhance primary productivity in a coastal basin. PLoS One 5 (10), e13255. https://doi.org/10.1371/journal.pone.0013255.

Roman, J., Estes, J.A., Morissette, L., Smith, C., Costa, D., McCarthy, J., Nation, J.B., Nicol, S., Pershing, A., Smetacek, V., 2014. Whales as marine ecosystem engineers. Front. Ecol. Environ. 12 (7), 377–385. https://doi.org/10.1890/130220.

Schoeman, R.P., Patterson-Abrolat, C., Plön, S., 2020. A global review of vessel collisions with marine animals. Front. Mar. Sci. 7, 292. https://doi.org/10.3389/fmars.2020.00292.

Singleton, B.E., Gillette, M.B., Burman, A., Green, C., 2021. Toward productive complicity: Applying 'traditional ecological knowledge' in environmental science. Anthropocene Rev. 1–22.

Smith, C.R., Glover, A.G., Treude, T., Higgs, N.D., Amon, D.J., 2015. Whale-fall ecosystems: recent insights into ecology, paleoecology, and evolution. Annu. Rev. Marine Sci. 7, 571–596.

Thomas, P.O., Reeves, R.R., Brownell Jr., R.L., 2016. Status of the world's baleen whales. Mar. Mamm. Sci. 32, 682–734.
Van der Wal, J.E.M., Spottiswoode, C.N., Uomini, N.T., Cantor, M., Daura-Jorge, F.G., Afan, A.I., Cram, D.L., 2022. Safeguarding human-wildlife cooperation. Conserv. Lett., e12886. https://doi.org/10.1111/conl.12886.
van Waerebeek, K., Baker, A.N., Félix, F., Gedamke, J., Iniguez, M., Sanino, G.P., et al., 2007. Vessel collisions with small cetaceans worldwide and with large whales in the Southern Hemisphere, an initial assessment. Lat. Am. J. Aquat. Mamm. 6, 43–69. https://doi.org/10.5597/lajam00109.
White, T.I., 2021. Dolphins, captivity, and cruelty. In: Thomas, N. (Ed.), Animals and Business Ethics. Springer, London.
Whitehead, H., Rendell, L., 2015. The Cultural Lives of Whales and Dolphins. The University of Chicago Press. ISBN: 978-0-226-89531-4.
Wright, A.J., Kyhn, L.A., 2015. Practical management of cumulative anthropogenic impacts with working marine examples. Conserv. Biol. 1–8.
Würsig, B., 2020. From science only to science for conservation: a personal journey. Ethics Sci. Environ. Polit. 20, 25–32. https://doi.org/10.3354/esep00191.

CHAPTER TWO

Discovering marine biodiversity in the 21st century

Alex D. Rogers[a,b,*], Ward Appeltans[c], Jorge Assis[d], Lisa T. Ballance[e], Philippe Cury[f], Carlos Duarte[g], Fabio Favoretto[h], Lisa A. Hynes[b], Joy A. Kumagai[i], Catherine E. Lovelock[j], Patricia Miloslavich[k,l], Aidin Niamir[i], David Obura[m], Bethan C. O'Leary[n,o], Eva Ramirez-Llodra[a,b], Gabriel Reygondeau[p,q], Callum Roberts[n], Yvonne Sadovy[r], Oliver Steeds[b], Tracey Sutton[s], Derek P. Tittensor[t], Enriqueta Velarde[u], Lucy Woodall[b,v], and Octavio Aburto-Oropeza[w]

[a]REV Ocean, Lysaker, Norway
[b]Nekton Foundation, Begbroke Science Park, Oxford, United Kingdom
[c]Intergovernmental Oceanographic Commission of UNESCO, Oostende, Belgium
[d]Centre of Marine Sciences, University of Algarve, Faro, Portugal
[e]Marine Mammal Institute, Oregon State University, Newport, OR, United States
[f]MARBEC, IRD, Univ. Montpellier, Brussels, Belgium
[g]King Abdullah University of Science and Technology (KAUST), Red Sea Research Center (RSRC) and Computational Bioscience Research Center (CBRC), Thuwal, Kingdom of Saudi Arabia
[h]Autonomous University of Baja California Sur, La Paz, Baja California Sur, Mexico
[i]Senckenberg Biodiversity and Climate Research Institute, Frankfurt am Main, Germany
[j]School of Biological Sciences, The University of Queensland, St Lucia, QLD, Australia
[k]Scientific Committee on Oceanic Research (SCOR), College of Earth, Ocean and Environment, University of Delaware, Newark, DE, United States
[l]Departamento de Estudios Ambientales, Universidad Simón Bolívar, Venezuela & Scientific Committee for Oceanic Research (SCOR), Newark, DE, United States
[m]CORDIO East Africa, Mombasa, Kenya
[n]Centre for Ecology & Conservation, College of Life and Environmental Sciences, University of Exeter, Penryn, United Kingdom
[o]Department of Environment and Geography, University of York, York, United Kingdom
[p]Yale Center for Biodiversity Movement and Global Change, Yale University, New Haven, CT, United States
[q]Nippon Foundation-Nereus Program, Institute for the Oceans and Fisheries, University of British Columbia, Vancouver, BC, Canada
[r]School of Biological Sciences, Swire Institute of Marine Science, The University of Hong Kong, Hong Kong
[s]Nova Southeastern University, Halmos College of Natural Sciences and Oceanography, Dania Beach, FL, United States
[t]Department of Biology, Dalhousie University, Halifax, NS, Canada
[u]Instituto de Ciencias Marinas y Pesquerías, Universidad Veracruzana, Veracruz, Mexico
[v]Department of Zoology, University of Oxford, Oxford, United Kingdom
[w]Scripps Institution of Oceanography, UC San Diego, San Diego, CA, United States
*Corresponding author: e-mail address: alex.rogers@revocean.org

Contents

1. Marine biodiversity: What is known? 24
   1.1 Introduction: The challenge 24
   1.2 Species richness and other concepts in marine biodiversity research 25
   1.3 Global marine biodiversity estimates 28

Advances in Marine Biology, Volume 93
ISSN 0065-2881
https://doi.org/10.1016/bs.amb.2022.09.002

1.4 The distribution of life in the ocean 31
1.5 Contributions to people from marine biodiversity 36
1.6 The consequences of human activities on ocean biodiversity 39
2. International legislation regarding marine biodiversity and its implementation 44
3. How marine biodiversity is studied today 47
3.1 Biodiversity and biodiversity science 47
3.2 Taxonomy 49
3.3 The future of taxonomic research 55
4. Conclusion: Integrated approaches 93
Acknowledgements 96
References 96

Abstract

We review the current knowledge of the biodiversity of the ocean as well as the levels of decline and threat for species and habitats. The lack of understanding of the distribution of life in the ocean is identified as a significant barrier to restoring its biodiversity and health. We explore why the science of taxonomy has failed to deliver knowledge of what species are present in the ocean, how they are distributed and how they are responding to global and regional to local anthropogenic pressures. This failure prevents nations from meeting their international commitments to conserve marine biodiversity with the results that investment in taxonomy has declined in many countries. We explore a range of new technologies and approaches for discovery of marine species and their detection and monitoring. These include: imaging methods, molecular approaches, active and passive acoustics, the use of interconnected databases and citizen science. Whilst no one method is suitable for discovering or detecting all groups of organisms many are complementary and have been combined to give a more complete picture of biodiversity in marine ecosystems. We conclude that integrated approaches represent the best way forwards for accelerating species discovery, description and biodiversity assessment. Examples of integrated taxonomic approaches are identified from terrestrial ecosystems. Such integrated taxonomic approaches require the adoption of cybertaxonomy approaches and will be boosted by new autonomous sampling platforms and development of machine-speed exchange of digital information between databases.

1. Marine biodiversity: What is known?

1.1 Introduction: The challenge

The ocean is by far the largest living space on Earth. It comprises 1.3 billion km^3 of seawater and 361 million km^2 of seafloor covering 71% of the Earth's surface. Life occurs throughout the ocean, in the deepest trenches (e.g. Philippines Trench and Challenger Deep in the Marianas Trench; Jamieson, 2015), and most extreme environments (e.g. around high temperature deep-sea hydrothermal vents; Van Dover, 2000) and even in the absence of oxygen (Danovaro et al., 2010; Mentel and Martin, 2010). Marine life has evolved over nearly 4 billion years (e.g. Tashiro et al., 2017) and its diversity provides numerous benefits to humankind (Worm et al., 2006), many of which we are only just uncovering. Understanding life in the ocean has extended our ideas about where and how life can exist in the rest of the universe and given rise to the new scientific discipline of astrobiology (Brack et al., 2001; Rampeletto, 2010). Yet, we still only have a primitive understanding of how life is distributed in the ocean, what molecules and metabolic processes it supports and how it contributes to the functioning of marine, and ultimately the Earth's systems and how it underpins human society and the well-being of its citizens. Sixty-six to ninety percent of marine life remains undescribed, and many geographic regions of the ocean as well as entire ecosystems remain poorly explored (Appeltans et al., 2012; Mora et al., 2011; Webb et al., 2010). As a result, human society cannot fully comprehend the vital importance of a healthy ocean as a critical part of Earth's life support system and hence how it contributes to their own well-being. This situation, however, has been accepted as the *status quo* with the consequences that much of the ocean has fallen out of sight and out of mind, and hence there has been systematic unsustainable exploitation of marine resources (e.g. overfishing), destruction of marine ecosystems (e.g. coastal development, pollution, trawling) and subsequent degradation of the ocean systems. The discovery of marine life and understanding where it is found in the ocean as well as how it changes in response to human impacts and interventions is a foundation to addressing the biodiversity crisis in the ocean and managing its recovery. Here we summarise what is known about the distribution of biodiversity in the ocean, why it is poorly understood and what technologies are now available or are required to meet the challenge of accelerating species discovery and monitoring of ocean life.

1.2 Species richness and other concepts in marine biodiversity research

A healthy ocean is one in which ecosystems host diverse life (Rogers et al., 2020), which can be quantified through metrics such as species richness to describe and measure this diversity (see below). Species richness is defined as the count of species within an ecosystem, habitat or sampling unit within a scientific study. It is the simplest measure of diversity as unlike many others it does not include consideration of species abundance or dominance (e.g.

Shannon-Weiner diversity measure). However, species richness does not include consideration of the evolutionary history of organisms in a sample and biodiversity measures that include phylogenetic information, such as average taxonomic distinctness also exist (e.g. Schweiger et al., 2008). Scientists and conservationists have also wanted to consider the taxonomic distinctness and threat status of species when planning management strategies for which measures of evolutionary distinctiveness and global endangerment (EDGE) have been developed (e.g. Collen et al., 2011). Here we mainly consider patterns of species richness in the ocean acknowledging that other measures of marine biodiversity are relevant in studies of biodiversity.

Another important concept in marine biodiversity research is that by which species themselves are defined. These concepts reflect the different evolutionary processes through which species arise. Probably the best-known species concept is the Biological Species Concept (BSC) which is defined as: *groups of actually or potentially interbreeding natural populations, which are reproductively isolated from other such groups* (Mayr, 1942, 1963). It has been known for some time, however, that this definition does not cater for the many situations where reproductive barriers are semi-permeable or where speciation can occur even where interbreeding is still taking place (e.g. Hausdorf, 2011). It also assumes biparental inheritance when some organisms exhibit uniparental inheritance (Hausdorf, 2011). It is also hardly applicable to organisms lacking sexual reproduction and proliferating through clonal processes, such as cell division in most unicellular microbes. One concept which has been applied to try and include such exceptions is the phylogenetic species concept (PSC) defined as: *a diagnosable cluster of individuals within which there is a parental pattern of ancestry and descent, beyond which there is not, and which exhibits a pattern of phylogenetic ancestry and descendent among units of like kind* (Eldredge and Cracraft, 1980). This concept answers some of the criticisms towards the BSC, but also does not fully encompass the range of variation in evolutionary mechanisms leading to speciation seen in nature (Hausdorf, 2011). Up to 24 species concepts were recognised by the late 1990s and further, definitions for these concepts can differ between scientists (e.g. De Queiroz, 2007). For the purposes of practicality this paper is agnostic towards species concepts, but it should be recognised that they can make a significant difference in the quantitative estimation of biodiversity (e.g. Agapow et al., 2004).

Three other terms related to species are also relevant when considering biodiversity: cryptic species, umbrella species and keystone species. Cryptic species are organisms where speciation has taken place with little or no recognisable morphological change (e.g. Fišer et al., 2018). Multiple evolutionary processes can be responsible for cryptic speciation in different lineages so there is no one explanation for this phenomenon (e.g. Fišer et al., 2018). Cryptic species are commonly encountered in the ocean and can lead to a serious underestimation of biodiversity (e.g. Fišer et al., 2018). The advent of molecular approaches to taxonomy or estimation of biodiversity has accelerated the recognition of these species in the environment. In many cases they remain undescribed in the formal sense but instead are referred to as molecular operational taxonomic units (MOTUs; e.g. Fišer et al., 2018) or, more recently, sequence variants, when assessments are based on nucleic acid sequencing (Porter and Hajibabaei, 2018). Whilst some authors reject the notion of the cryptic species concept arguing that such species often do have morphological and/or ecological differences when examined at a more detailed level (e.g. Korshunova et al., 2019), there is no doubt that such organisms pose both a challenge and an opportunity for taxonomists and biodiversity scientists. The challenge is that the rapid discovery of such organisms is leading to a large backlog of M-OTUs with no formal species descriptions. The opportunity is that molecular approaches (e.g. molecular barcoding or metabarcoding such as environmental-DNA [eDNA] approaches) may offer new ways of better understanding the true diversity within ecosystems and how it relates to ecosystem function, as well as dissociate progress in biodiversity research from the requirements of formal taxonomic classification.

The umbrella species concept is a term used in conservation science and is based on the premise that by focusing on the requirements of particularly demanding species encapsulates the demands of many co-occurring, less demanding species (e.g. Roberge and Angelstam, 2004). Thus, by focusing conservation efforts on the most demanding species it is possible to also conserve co-occurring species (Roberge and Angelstam, 2004). This concept will not be discussed further here but may be of relevance when considering the management implications of biodiversity studies aimed at reversing species declines.

Organisms exist embedded in complex networks that delineate interdependencies on one another of the same and different species. Those that occupy the central positions of these networks thereby connected with many interactions with other members of biological communities have been termed keystone species, and can be expected to have important roles in structuring an entire ecosystem. An example is the NE Pacific where sea otters predate on sea urchins which graze on kelp. By controlling urchin numbers, sea otters permit the kelp forest to flourish but when the otters are removed by hunting for example the urchins explode in numbers and decimate the kelp forests (e.g. an example of top-down control; Estes and Palmisano, 1974; Estes and Duggins, 1995). An example from pelagic ecosystems is where small fish exert a major control on the trophic dynamics of upwelling ecosystems and constitute midtrophic-level 'wasp-waist' populations. Ecosystem dynamics can be entirely different depending on how the food web is controlled (bottom-up, top-down or wasp-waist; Cury et al., 2000).

Documenting and understanding keystone species is valuable for guiding ocean management and is an example of where not all species are equal in management terms.

1.3 Global marine biodiversity estimates

Estimates of global metazoan marine species diversity over the last 10 years vary depending on the methods used from approximately 300,000 to 2.2 million (Appeltans et al., 2012; Costello et al., 2012; Mora et al., 2011). This is significantly lower than that on land (e.g. May, 1988, 1994; Mora et al., 2011) which is dominated by insects now estimated to number in the region of 5.5 million species (Stork et al., 2015). However, although species richness is higher on land the diversity of higher taxonomic categories such as phyla and orders is much higher in the ocean (e.g. May, 1994). The World Register of Marine Species (https://www.marinespecies.org/index.php), for example records 33 animal phyla of which 32 are found in the ocean and only 17 in terrestrial and freshwater ecosystems (Table 1). Only one phylum in this list does not occur in the ocean although we note that the phylum Onycophora (velvet worms) is missing from this list and this is also only found on land.

The high species richness in terrestrial ecosystems reflects a higher speciation rate in organisms on land which has been explained through greater barriers to dispersal and increased rates of allopatric speciation (Wiens, 2015). Plant-pollinator co-evolution also explains the large number of insect species whereas flowering plants are scarce in the marine environment (Hemminga and Duarte, 2000). However, note that high extinction rates and slow recovery from extinctions for marine taxa may have also played a significant role in the contrast in species richness between the land and the ocean (Miller and Wiens, 2017). Higher phyletic diversity in the ocean than that on land or in freshwater has been suggested to reflect the length of time life has inhabited the ocean and that seawater provides a relatively

Table 1 Number of animal species distributed across phyla in the ocean and on land. These are accepted species names. Data from the WORMs Register of Marine Species (https://www.marinespecies.org/index.php)

Taxon	Number of marine species	Number of non-marine species
Animals		
Acanthocephala	512	9
Annelida	13,341	741
Arthropoda	57,271	29,937
Brachiopoda	414	0
Bryozoa	6,330	104
Chaetognatha	132	0
Chordata	23,838	15,526
Cnidaria	11,952	27
Ctenophora	203	0
Cycliophora	2	0
Dycemida	122	0
Echinodermata	7,478	0
Entoprocta	197	2
Gastrotricha	515	346
Gnathifera	0	1
Gnathostomulida	100	0
Hemichordata	132	0
Kinorhyncha	307	0
Loricifera	29	0
Mollusca	49,788	32,240
Nematoda	6,406	4,886
Nematomorpha	5	0
Nemertea	1,316	13
Orthonectida	25	0
Phoronida	13	0

Continued

Table 1 Number of animal species distributed across phyla in the ocean and on land. These are accepted species names. Data from the WORMs Register of Marine Species (https://www.marinespecies.org/index.php)—cont'd

Taxon	Number of marine species	Number of non-marine species
Placozoa	3	0
Platyhelminthes	12,884	2,958
Porifera	9,126	232
Priapulida	22	0
Rotifera	182	117
Sipuncula	158	0
Tardigrada	212	1,019
Xenacoelomorpha	452	2

benign medium that can support a wide range of body plans (Brusca and Brusca, 1990; May, 1994). Looking at global living biomass on Earth gives a different perspective to the distribution of biodiversity. Plants make up the vast majority of biomass on Earth (~450 Gt) and are primarily located on land (Bar-On et al., 2018). However, if animals alone are considered, a very different pattern is observed. Marine arthropods are estimated to have the greatest biomass of any group of animals on Earth (~1 Gt) with groups such as copepods and even single species, such as Antarctic krill, *Euphausia superba* (~0.05 Gt), making a significant contribution to global animal biomass (Bar-On et al., 2018). The biomass of terrestrial arthropods, including the insects (0.2 Gt) is estimated to be considerably smaller than that of marine invertebrates (Bar-On et al., 2018). The fishes are the second largest group of animals on Earth in terms of biomass (0.7 Gt), being dominated by the mesopelagic fish (those living between 200 and 1000 m depth in the ocean; Irigoien et al., 2014; Bar-On et al., 2018).

Examination of the patterns of described species vs estimated global species richness in the ocean reveals that, on the whole, vertebrate groups are well described with 77% (Pisces including Agnatha) to an estimated 100% (Carnivora and Sirenia; Appeltans et al., 2012; see also Scheffers et al., 2012) of taxa described. Marine invertebrates are much less studied, and their diversity can be much higher than vertebrate groups with as few as 3% of the estimated number of species in the ocean described (e.g.

Loricifera and Tantulocarida; Appeltans et al., 2012). Within the invertebrates, taxa with relatively few species (e.g. the crustacean groups Lomisoidea with 1 species or the Euphausiacea with 86 species) or those which are relatively large and obvious in the marine environment (e.g. Echinodermata) are relatively well described (Appeltans et al., 2012). Groups with fewer than 20% of species described include a wide range of taxa with a small body size including the Cycliophora, Loricifera, various groups of peracarid crustaceans and the Nematoda (Appeltans et al., 2012). For these groups typically hundreds of species remain undescribed though for some this runs into the tens of thousands or even more than a hundred thousand including the Isopoda (63,150–123,600 species), the Gastropoda (85,000–105,000 species) and Nematoda (50,000 species; Appeltans et al., 2012; Bouchet et al., 2016). Invertebrate groups with relatively few morphological characters, such as the Hydrozoa and Nemertea are prone to cryptic speciation (species only distinguishable using molecular methods), creating further problems in distinguishing taxa (e.g. Krämer et al., 2017; Moura et al., 2008). Nematoda are often parasitic and, therefore, not readily visible, unless potential host animals are dissected, which may also render them to be cryptic and poorly resolved (Costello, 2016; Costello and Chaudhary, 2017).

1.4 The distribution of life in the ocean

Knowledge of the patterns of biodiversity in the ocean are based on highly biased sampling (Rogers et al., 2020). Understanding of marine biodiversity varies markedly across regional, national and, more importantly, trophic levels (Costello et al., 2010). Data from the Census of Marine Life (CoML) programme, a 10-year global exploratory assessment of marine biodiversity, are available in the Ocean Biodiversity Information System (OBIS) of the Intergovernmental Oceanographic Commission (IOC) of the United Nations Educational, Scientific and Cultural Organization (UNESCO). The CoML data suggest that, in relative terms, marine species within Chinese, Australian and European waters are best known with the tropical western Atlantic, tropical eastern Pacific and Canadian Arctic regions being poorly studied (Costello et al., 2010). Ecosystems that are particularly poorly known include the deep sea, coral reefs, ice-covered areas and chemosynthetic habitats (Costello et al., 2010; Webb et al., 2010). In general terms, as you move away from the coast and into deeper water there are less data (e.g. Webb et al., 2010). Knowledge of the identity and distribution of commercially exploited taxa is greater than that of non-extracted taxa, and larger organisms tend to be better known than smaller ones (Fautin et al., 2010; Worm and Tittensor, 2018). Currently, only a relatively small number of species are considered to have enough independent records that describe their full geographic distribution (about 50,000 species; Gagné et al., 2020). Species lacking such substantial spatial information are likely to have large uncertainties around their true distribution. Emblematic (mammals, corals or fish) and exploited species (fish and invertebrates) are among the most well-documented spatially.

Geographic areas with the best knowledge of marine biodiversity do not match well with areas of highest diversity, reflecting the distribution of both historical and present-day scientific capacity for taxonomy. Historically, highly sampled regions are often located in the Northern Hemisphere in the coastal regions around countries with high-income economies where research institutions abound. It is crucial to account for such sampling bias when examining the distribution of biodiversity (Gagné et al., 2020; Hughes et al., 2021; Tittensor et al., 2010). At a global scale analyses of marine biodiversity data indicate a peak at tropical to sub-tropical latitudes with particular hotspots focused around the Indo-Pacific Coral Triangle and to a lesser extent the Caribbean (Asch et al., 2018; Reygondeau, 2019; Rogers et al., 2020; Tittensor et al., 2010; Figs 1 and 2). The central and western Indian Ocean, Red Sea, South West Pacific Islands (i.e., the Bismarck Archipelago, the Great Sea Reef of Fiji, New Caledonia, New Guinea, the Solomon Islands, Vanuatu) and Southeast Asia also show the highest levels of species richness (Rogers et al., 2020; Figs 1 and 2). Explanations for the low latitude peak in biodiversity include: speciation and extinction rates over geological timescales as correlated with latitude (Crame and Clarke, 1997; Jablonski et al., 2006, 2013); seafloor dynamics, including the breakup of the Tethys Ocean (Hou and Li, 2018); ecological drivers such as habitat area, land vs ocean area by latitude, sea surface temperature (Worm and Tittensor, 2018), and intrinsic biological traits such as larval development mode and interspecies interactions (Edgar et al., 2017a; Pappalardo and Fernández, 2014; Roy et al., 1998). There are also steep longitudinal gradients in diversity, with an increase from both east and west towards Southeast Asia, and from east to west in the tropical Atlantic (Rogers et al., 2020; Fig. 1). Whilst some large groups of organisms follow the general pattern of a tropical/sub-tropical peak in species richness (e.g. reef-building corals, coastal fishes, shallow-water ophiuroids, cone snails, mangroves, coastal cephalopods, lobsters and gastropods) others do not (Rogers et al., 2020).

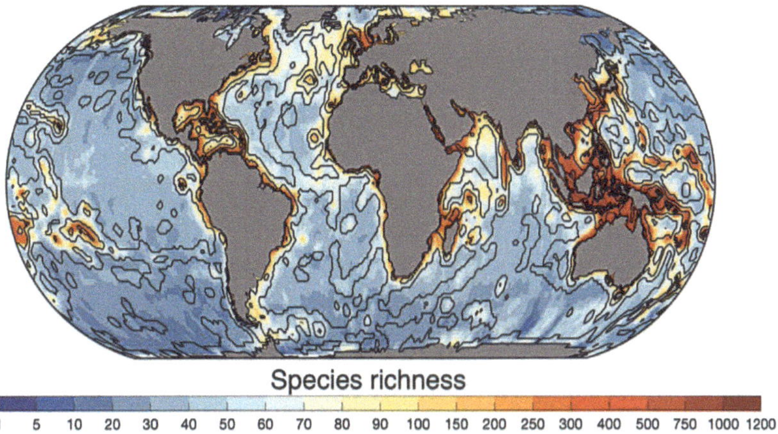

Fig. 1 Pattern of global species richness across the global ocean based on records from the Ocean Biodiversity Information System (OBIS), the Global Biodiversity Information Facility (GBIF), Fishbase, the Coastal and Oceanic Plankton Ecology Production and Observation Database (COPEPOD), the Jellyfish Database Initiative and the International Union for Conservation of Nature (IUCN). Data were filtered according to the method in Gagné et al. (2020) and the study. *From Rogers, A., Aburto-Oropeza, O., Appeltans, W., et al., 2020. Blue Paper 10: Critical Habitats and Biodiversity: Inventory, Thresholds, and Governance. Report prepared for the (Norwegian) Prime Minister's High-Level Panel on a Sustainable Ocean Economy. 85 pp. https://www.oceanpanel.org/blue-papers/critical-habitats-and-biodiversity-inventory-thresholds-and-governance.*

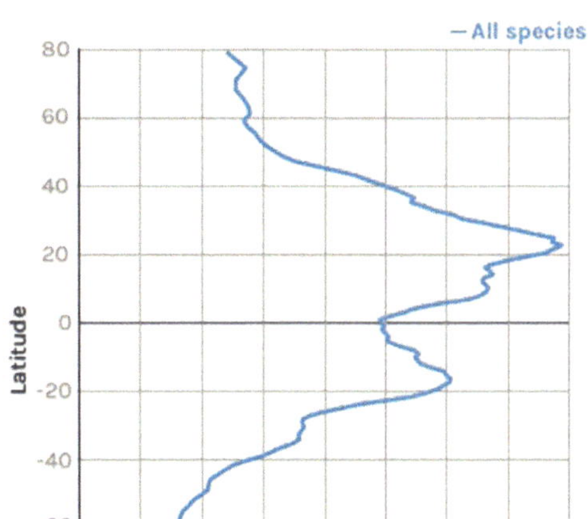

Fig. 2 Latitudinal pattern of species richness based on the same dataset as Fig. 1. *From Rogers, A., Aburto-Oropeza, O., Appeltans, W., et al., 2020. Blue Paper 10: Critical Habitats and Biodiversity: Inventory, Thresholds, and Governance. Report prepared for the (Norwegian) Prime Minister's High-Level Panel on a Sustainable Ocean Economy. 85 pp. https://www.oceanpanel.org/blue-papers/critical-habitats-and-biodiversity-inventory-thresholds-and-governance.*

Seagrass and seaweeds tend towards a more temperate centre of the distribution of diversity probably because of availability of macronutrients and conditions of temperature (Gaines and Lubchenco, 1982; Jayathilake and Costello, 2021; Keith et al., 2014; Kerswell, 2006; McKenzie et al., 2020; Short et al., 2007; Tittensor et al., 2010; Worm and Tittensor, 2018) whilst seals show an opposite pattern to the normal by having a peak in diversity in sub-polar and polar environments (Pompa et al., 2011; Tittensor et al., 2010). This is likely because warm-blooded organisms have a metabolic advantage at lower temperatures (endothermy; Grady et al., 2019; Pompa et al., 2011).

Biodiversity in the open ocean shows a generally bimodal pattern (Chaudhary et al., 2016), with pelagic zooplankton such as foraminifera, copepods and euphausiids, open ocean fishes such as tuna and billfishes, pelagic sharks, and cetaceans all showing a mid-latitudinal peak in species richness, generally between latitudes 30 and 40 degrees (Tittensor et al., 2010). Some differences between these taxa are apparent, including cetaceans being widely distributed in terms of richness peaks across latitudinal bands, whereas pelagic shark hotspots tend to skew towards the coast. Pelagic cephalopods are under-sampled, but they appear to show a similar intermediate latitudinal peak, albeit only in the Northern Hemisphere (Tittensor et al., 2010). Pelagic seabirds (such as albatross and petrels) show a mid-latitude peak, but only in the Southern Hemisphere (Davies et al., 2010).

Global patterns of species richness in the deep sea are poorly understood, in part because of the sampling challenges. A study on a single group of organisms, the brittle stars (Ophiuroidea) has shown that in shallow water they have a peak in diversity at low latitudes on the shelf and slope but in waters deeper than 2,000 m they show peak richness at temperate latitudes (30°–50°; Woolley et al., 2016). This pattern of richness seems to reflect the input of particulate organic matter into the deep ocean, the main food source for bathyal and abyssal organisms.

Other patterns of biodiversity, including intraspecific genetic variation and habitat diversity, are also not well described (Blasiak et al., 2020; Fautin et al., 2010), with some exceptions. Global habitat diversity shows a similar pattern to that of species richness (Rogers et al., 2020; Fig. 3). This is, however, based on a limited number of habitats which were selected for both data availability and their ecological and socioeconomic importance (Rogers et al., 2020). These habitats were: estuaries, mangrove forests, saltmarsh, seagrasses, coral reefs, macroalgal forests shelf valleys and canyons,

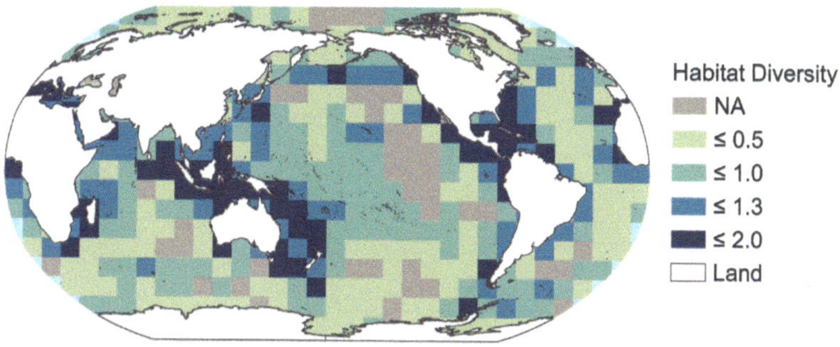

Fig. 3 Global habitat diversity based on Shannon Weiner Diversity Index (Rogers et al., 2020). Data were drawn from: Alder (2003; updated by UNEP-WCMC); Giri et al. (2011; updated by UNEP-WCMC); McOwen et al. (2017; updated by UNEP WCMC); Short (2018; and UNEP WCMC); UNEP-WCMC, WorldFish Centre, WRI, TNC. (2018); Harris et al. (2014); Freiwald et al. (2017; updated by UNEP WCMC); Beaulieu and Szafranski (2018).

cold-water coral reefs, seamounts, trenches, hydrothermal vents and ridges (Rogers et al., 2020). Habitat diversity appears higher nearer the coast reflecting the coastal and shelf ecosystems selected in the study (Rogers et al., 2020; Fig. 3). However, knowledge of distribution of deep-sea habitats is relatively poor (Rogers et al., 2020). The latitudinal distribution of habitats based on this limited data set also shows a low latitude peak in mean number of habitats per cell but the pattern is not as clear as for species richness, especially in the northern hemisphere (Rogers et al., 2020; Fig. 4). There are limitations to these data in terms of number of habitats represented and in sampling bias across latitude.

The account above is focussed on metazoans, whereas microbes are dominant components of the ocean ecosystem, in functional terms. However, these organisms cannot be readily included in biodiversity inventories, both because of insufficient exploration and also difficulties in applying standard concepts of species definition to these organisms. Novel discoveries using molecular assessments (see below) are rapidly increasing our understanding of the diversity and distribution of microbes in the ocean, both through amplicon sequencing and, with greater detail, metagenomic approaches that capture both biodiversity and metabolic functions across microbial communities (Obiol et al., 2020; Salazar and Sunagawa, 2017). For example, global microbial taxonomic richness in marine sediments predicted by species–area relationship models is 7.85×10^3 to 6.10×10^5 for Archaea and 3.28×10^4 to 2.46×10^6 for Bacteria as amplicon sequence variants, comparable to the richness in seawater and that in topsoil,

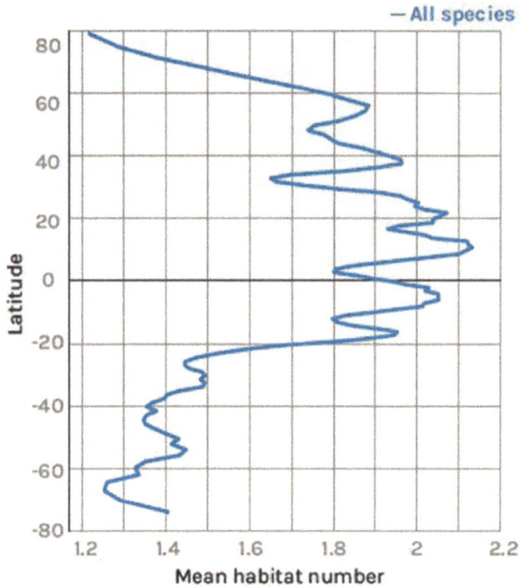

Fig. 4 Latitudinal pattern in mean number of habitats per geographic cell based on the data in Fig. 3 (Rogers et al., 2020).

with an important role for oxygen presence or absence and organic carbon concentration in driving the distribution of benthic microbes (Hoshino et al., 2020). Microbial diversity patterns are emerging from these efforts, pointing at microbial diversity peaks at latitudes 30° north and south, with consistently greater richness predictions at higher latitudes compared with lower ones (Ladau et al., 2013).

1.5 Contributions to people from marine biodiversity

The biosphere, the part of the Earth system where life is found, is a self-organising regenerative entity where organisms make use of the non-living or abiotic resources such as light, carbon dioxide and macronutrients and transform them (Dasgupta, 2021). Production of food by natural ecosystems is a result of such transformations that directly benefits humans whilst the water, carbon and other nutrient cycles are aspects which indirectly benefit us. Such benefits are known as Ecosystem Services although more lately the term Nature's Contributions to People (NCP) has been used. The concept of NCP incorporates not only positive benefits but also negative consequences arising from natural processes (e.g. harmful diseases transmitted by vectors such as mosquitoes). It also broadens ecosystem services from ecology and economics to incorporate social science perspectives and indigenous and local knowledge systems de-emphasizing their economic contribution as a criteria to assign importance (Díaz et al., 2018). The Common International Classification of Ecosystem Services (CICES) categorised ecosystem services into:

- Provisioning services
  - The products of ecosystems on which we rely such as food, fresh water, fuel (dung, wood, twigs and leaves), fibre (grasses, timber, cotton, wool, silk), biochemicals and pharmaceuticals (medicines, food additives), genetic resources (genes and genetic information used for breeding of organisms and biotechnology) and ornamental resources (skins, shells, flowers).
- Regulating and maintaining services
  - The ways in which the biosphere regulates and maintains ecosystem processes such as climate, atmospheric composition and the supply of water.
- Cultural services
  - Non-material benefits such as spiritual experiences, religious practice, inspiration and education (Dasgupta, 2021).

The newer framework for NCP is summarised as (Díaz et al., 2018):

- Material Nature's Contributions to People
  - Substances, objects, or other material elements from nature that directly sustain people's physical existence and material assets (e.g. food, medicinal and genetic resources).
- Regulating Nature's Contributions to People
  - Functional and structural aspects of organisms and ecosystems that modify environmental conditions experienced by people and/or regulate the generation of material and non-material contributions (e.g. habitat provision and maintenance, regulation of ocean acidification).
- Non-Material Nature's Contributions to People
  - Nature's effects on subjective or psychological aspects underpinning people's quality of life, both individually and collectively (e.g. supporting identities, providing physical and psychological experiences).

Biodiversity underpins ecosystem function and the provision of nature's contributions to people. Although the exact relationship between biodiversity and ecosystem function is unclear (Rogers et al., 2020; Fig. 5), and may vary at a range of scales and with different species, it is generally accepted that loss of biodiversity leads to a degradation of ecosystem function and this has an impact on NCP (e.g. Dasgupta, 2021). In economic terms, ecosystems

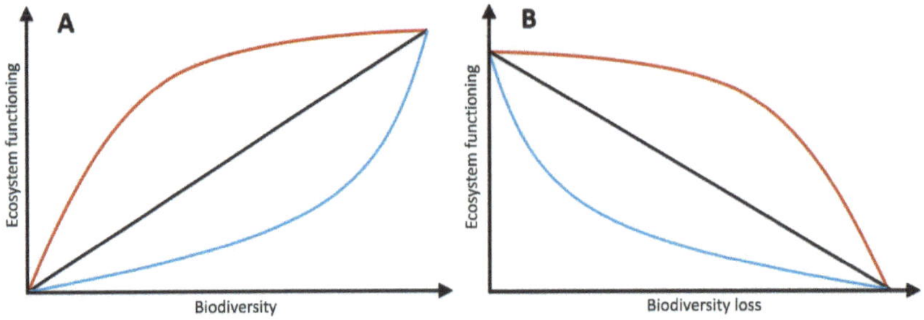

Fig. 5 Possible shapes of the Biodiversity Ecosystem Function (BEF) curve (left) and possible responses of communities to decreasing biodiversity in terms of loss of ecosystem services (right; modified from Naeem, 2002; Strong et al., 2015). The red curve indicates a situation where there is a considerable level of functional redundancy between species leading to significant resilience to biodiversity loss. The black curve indicates a situation where ecosystem function is directly correlated to biodiversity. The blue curve represents a situation of accelerating ecosystem function with species diversity (e.g. as a result of complementarity between species). In such a situation a small loss in diversity leads to a large loss in ecosystem function.

can be viewed as capital goods (natural capital or natural assets) which depreciate in value when they are degraded through biodiversity loss (Dasgupta, 2021). On the other hand, if ecosystems are sustainably used, they will regenerate through natural processes mediated by biodiversity which is also considered to increase long-term ecosystem resilience to pressures (Oliver et al., 2015). From the more holistic view of sustainable use of nature put forward by IPBES, nature can be seen to also include humans, instead of being diametrically opposed or in a hierarchical relationship. Therefore, a good quality of life can be reconciled with that of life on Earth advancing the intertwined economic, social and environmental dimensions of sustainable development in a balanced manner (IPBES, 2022).

Nature's contributions to people can be valued in many different ways. There have been various attempts over the past two decades to put a monetary value on NCP provided by the ocean (e.g. Costanza et al., 1997, 2014; Martin et al., 2016; WWF, 2015). For example, WWF (2015) estimated that ocean assets were worth \$24 trillion and produce goods and services worth \$2.5 trillion per year. However, there are many issues with such valuations as they cannot be readily translated to monetary value in real economies. For example, whilst it may be relatively straightforward to estimate the value of Material NCP, such as for marine capture fisheries (e.g. \$109 billion; Sumaila et al., 2015), estimating the value for Regulating NCP is much more difficult to the point where they can be largely invisible (Dasgupta, 2021).

This has contributed to the lack of impact of purely economic valuations of NCP on policy related to use of natural resources and ultimately biodiversity loss (Díaz et al., 2018). An exception to this has been the development of carbon credit markets that now include ecosystems processes, extending from the climate regulation services of forests to those of so-called Blue Carbon habitats such as mangrove forests, seagrass meadows and salt-marshes (Nelleman et al., 2009). Blue carbon habitats rank among the most intense carbon sinks in the biosphere (Duarte et al., 2013), and such climate services can now be assigned a monetary value under carbon credit schemes. Other issues include that compared to the land where the extent of habitats or ecosystems can easily be surveyed for example by satellite remote sensing (e.g. forests) the ocean is a three-dimensional system, again leading to a problem of invisibility (Rogers et al., 2020). Competing ecosystem services, especially those that are easily monetised such as Material NCPs, mean that these 'invisible' contributions are often degraded at the cost of maximising extraction of desired goods from nature (Dasgupta, 2021). Many natural assets are also public goods or the product of common assets leading to further problems of valuation and unsustainable exploitation (Rogers et al., 2020). A good example of this is shrimp aquaculture in SE Asia, where mangrove forests are destroyed to build farms. Economic studies indicate that although shrimp farms are profitable to their investors, economically the activity can generate a loss through the destruction of the regulatory functions of the mangrove forest as well as its direct provision of goods to local villagers (the commoners) such as wood, honey and seafood (Sathirathai and Barbier, 2001). One of the largest benefits of the mangrove forests in terms of value in this case study was coastal protection, yet this value is not integrated into accounting frameworks. The shrimp/mangrove study also illustrates that ecosystem services are provided across a range of scales and can have both positive benefits and negative benefits to different elements of human society (Sathirathai and Barbier, 2001). It is important to recognise that maximising production for goods that we want, often by replacing natural processes with artificial ones (e.g. creation of shrimp ponds converted from mangrove forests), degrades biodiversity and ecosystem function not just in the present but also for future generations (Dasgupta, 2021).

1.6 The consequences of human activities on ocean biodiversity

Although there is evidence of localised depletion of ocean resources even in ancient times, humans did not drive species to extinction as they did on land probably because marine species have wide geographic ranges or refugia from hunting and gathering (Erlandson and Rick, 2010; Hoffman, 2005). The first evidence of the large-scale collapse of fisheries in Europe occurred in the High Middle Ages through to the Late Middle Ages (Hoffman, 2005). During this period, religious practice encouraged the replacement of meat with seafood on up to 130 days of the year, driving demand for fish (Hoffman, 2005). At the same time, land-use changes caused significant alterations in freshwater and coastal ecosystems including sedimentation and eutrophication (Hoffman, 2005). The combination of fishing pressure and changes in water quality in freshwaters and estuaries drove a decline in these species causing them to disappear in many European waterways and coastal ecosystems by the 12th Century for sturgeon and the Late Middle Ages for salmon (Hoffman, 2005). As these inland and coastal fisheries resources declined and technologies in sailing, fishing and preservation of fish improved, attention turned to offshore marine fish stocks, especially cod and herring in the Baltic and North Seas (Hoffman, 2005). Trade in fish was one of the drivers of the medieval commercial revolution, and the wealth of the Hanseatic League, a powerful trading confederation of the time, was partially founded on herring (Cripps, 2014; Hoffman, 2005). However, by the Late Medieval Age a combination of heavy fishing, declining water quality and climate variation led to the successive collapse of important fish stocks including in the southern Baltic and North Seas (Hoffman, 2005). The extractive practices of Europeans were not necessarily reproduced by other cultures, which included aquaculture or fish ranching (Egypt and Asia), and traditional practices that promoted sustainable harvesting, such as temporary closures in Pacific Islands (Veitayaki, 1997), and habitat enhancement practices embedded in traditional 'satoumi' practices of Japanese fishers (Berque and Matsuda, 2013).

The first known marine extinctions caused by humans occurred during colonial times, an example being the elimination of Stellar's sea cow in the North Pacific in 1768 (Jackson et al., 2001). The geographic distribution of this species was probably limited as a result of recent glaciations but may have been further reduced by Aboriginal hunting (Jackson et al., 2001). Stellar's sea cow was discovered when the *St Peter*, Vitus Bering's ship leading the Great Northern Expedition, was wrecked in 1741 on Bering Island in the Commander Islands group. Within 27 years of its discovery, it was extinct as a result of hunting for its meat, fat and hide. The Caribbean monk seal (*Monachus tropicalis*) was heavily exploited from the 17th century, although records of hunting go back as far as the 15th century (McClenachan and

Cooper, 2008). Its exploitation, which was for food and oil, used in machinery for sugar processing, expanded with the colonisation of Caribbean Islands by European powers, although the species did not go extinct until 1952 (McClenachan and Cooper, 2008).

Whilst there have been relatively few documented cases of marine extinctions there is ample evidence of range contractions and large impacts on species abundance and ecosystem structure as a result of human activities (e.g. Dulvy et al., 2003; McCauley et al., 2015). During the colonial period, large-scale exploitation of fish and shellfish stocks in North America and as far away as Australia led to serial depletion of stocks (e.g. Kirby, 2004; Roberts, 2007). The destruction of oyster beds was completed when technological advances allowed the development of mechanical methods for harvesting in the late 19th century (e.g. Beck et al., 2011). The industrial revolution led to the development of steam trawlers in the 1880s. Steam trawlers could fish further offshore for longer periods of time and deploy larger gear in deeper waters than sailing vessels (Thurstan et al., 2010). As early as 1885, outcries from line fishers in the UK led to enquiries with regards to the damage to both fish stocks and marine ecosystems by trawling. Further technological advances in fishing, including larger and more powerful vessels, new types of fishing gear, freezer and canning technology, navigation and sonar equipment led to a large expansion of global fishing following World War II (a period known as the Great Acceleration; Standing, 2022) and subsequent depletion and collapse of some of the world's largest fish stocks as well as numerous populations of whales and other species. For an account of the overexploitation of fisheries for whales, seals, finfish and other species during the colonial period to recent times see Roberts (2007). Overexploitation of species targeted by fisheries has also led to widespread ecosystem impacts through cascading top-down effects and other consequences (e.g. Jackson et al., 2001; Myers et al., 2007; Baum and Worm, 2009). Overfishing and collapse of fish stocks also has severe socioeconomic consequences including impacts on livelihoods, food security and social stability (e.g. Pomeroy et al., 2016).

There have been various attempts to quantify losses to marine biodiversity. The Living Planet Index assesses changes in abundance in a range of species including mammals, birds, reptiles, amphibians and fish. These include terrestrial, freshwater and marine species (WWF, 2020). Overall, this indicates a loss of 68% in population abundance of the assessed groups since 1970 with decline continuing (WWF, 2020). Another approach to assess how biodiversity is being lost is the IUCN Red List, a tool which quantitatively estimates extinction risk in organisms (IUCN, 2017). Although, as for the LPI, this is a taxonomically highly biased dataset (Cowie et al., 2022), and only about 3% of described marine species have been assessed to date (Sullivan et al., 2019), it is one of few quantitative datasets to monitor biodiversity loss at the species level. Furthermore, even where groups are assessed many are classified as Data Deficient; in other words there are insufficient data to carry out an assessment of extinction risk (Cowie et al., 2022; Rogers et al., 2020). Based on 12 groups of marine invertebrates and vertebrates, Rogers et al. (2020) identified that the lower and upper bound for the proportion of species within individual taxa threatened with extinction ranges from 11% to 46% for marine invertebrates. For marine vertebrates only ∼4% of actinopterygians (ray-finned fish) are threatened with extinction. In contrast, the abundance of oceanic sharks and rays has declined by 71% since 1970 and now more than three quarters are threatened with extinction (Pacoureau et al., 2021).

An analysis of Red List Assessments indicates that by far the most significant driver of biodiversity loss in the ocean is overfishing and the destructive effects of fishing (e.g. habitat damage by trawling or bycatch of seabirds with longlines; Rogers et al., 2020). This is followed by coastal development, pollution, climate change, invasive species and transportation as drivers of extinction risk (Rogers et al., 2020). Similar findings arose from the Global Assessment Report on Biodiversity and Ecosystem Services (IPBES, 2019). Specific drivers of extinction risk differed in their importance amongst different taxonomic groups with fishing, for example, being the main driver of extinction risk for chondrichthyans, whilst invasive species were the main driver of seabird extinction risk (Rogers et al., 2020; Fig. 6). It is notable that the estimated proportion of marine species threatened with extinction (11–46%) spans that of species from terrestrial groups (20–25%; Webb and Mindel, 2015). Despite that, relatively few marine extinctions are known (Dulvy et al., 2003; McCauley et al., 2015). This could reflect the lower levels of observation and monitoring of ocean biodiversity which means that some extinctions may not been recorded. However, detailed studies of extinction levels in fish and molluscs suggest that marine taxa are genuinely at a lesser extinction risk than terrestrial or freshwater (Cowie et al., 2022; Miranda et al., 2022).

Another way of assessing biodiversity loss is through estimation of losses of habitat. This is easiest for coastal ecosystems, especially for those that can be surveyed through satellite remote sensing, such as mangrove forests. For example, seagrass beds have been estimated as losing 10% of their area

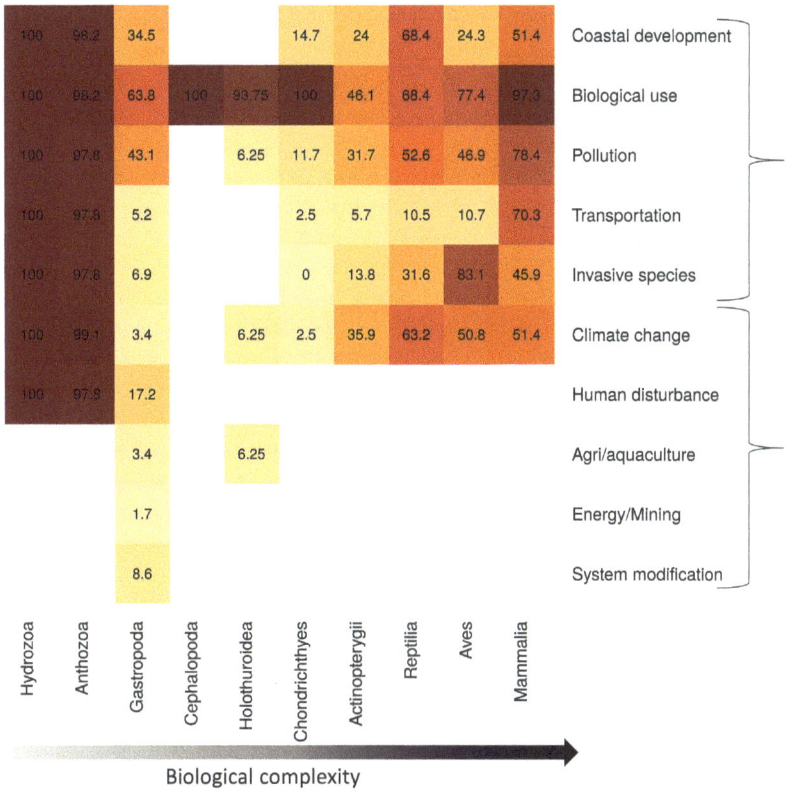

Fig. 6 Drivers of extinction risk in marine taxa assessed by Rogers et al. (2020). Overall, the area of the ocean remaining with no discernible human impact has been estimated at 3% (Halpern et al., 2015) whilst that which can be classified as wilderness (low human impacts) has been estimated at 13% (Jones et al., 2018a). An estimated 66% of the ocean is suffering increasing human impacts (Halpern et al., 2015).

globally per decade from 1970 to 2000 (Thomas et al., 2017). The global cover of mangroves has declined by about 40% (Thomas et al., 2017) and that of saltmarshes by about 60% (Gedan et al., 2009). However, loss rates of mangroves are slowing down globally and seagrass meadows are also expanding in some regions, particularly in the Northern Hemisphere (Duarte et al., 2020a). In contrast, coral reefs have lost about half of their coral cover since the 1870s with losses accelerating in the last two to three decades and projected to accelerate event further in the future, as a direct result of ocean warming (IPCC, 2019; Wilkinson et al., 2016a). Trends for macroalgal forests are more difficult to ascertain because they are highly dynamic ecosystems (Krumhansl et al., 2016; Wernberg et al., 2019).

Regionally significant losses of this habitat have been recorded (e.g. the Great Southern Reef of Australia; Bennett et al., 2016; Filbee-Dexter et al., 2020) whilst expansions are also recorded with the retreat of Arctic ice (Krause-Jensen et al., 2020), so globally kelp beds may not be losing area although they may lose species.

2. International legislation regarding marine biodiversity and its implementation

At least 23 international treaties, conventions and agreements relate to protection of marine biodiversity and/or the marine environment (Rogers et al., 2020; Table 2). This does not take into account regional and sub-regional agreements which impose further obligations on nations to protect the marine environment and biodiversity, in addition to national legislation. These treaties fall into three categories: biodiversity protection; fisheries regulation and regulation of activities such as navigation; prevention of pollution (Rogers et al., 2020). Most conventions are concerned with the management of marine living resources (11) or with preventing damage to the environment by pollution (10). A lesser number (8) are concerned with managing or conserving non-commercial species and fewer still deal with habitat protection (6; Rogers et al., 2020). Five conventions specifically require the initiation of marine protected areas (Rogers et al., 2020). Despite the number of conventions, as well as regional and sub-regional agreements and national legislation, biodiversity is still in decline in the ocean (Rogers et al., 2020). Several problems with international legislation have been identified which have allowed biodiversity to continue to be lost in the ocean. These are summarised from Rogers et al. (2020) as:

- Lack of effective implementation of treaties and conventions
- Sectorialisation of both conventions and their implementation agencies missing out on synergies and prone to unintended consequences.
- Prioritisation of commercial concerns (e.g. extraction of marine living resources) whilst there is a lack of prioritisation of biodiversity protection
- Lack of target setting to achieve the goals of conventions and treaties
- Lack of the ability to monitor trends in biodiversity
- Gaps in conventions and agreements, most notably for protection of biodiversity in areas beyond national jurisdiction

One of the significant barriers to governments implementing biodiversity treaties, conventions and agreements is the limited state of knowledge of

Table 2 Characteristics of the international conventions and agreements that have a bearing on protection of marine biodiversity or the marine environment.

Convention/agreement	A	B	C	D	E	F	G	H	I	J	K
1. IWC	Yes	Yes		Yes		Yes					
2.Convention on fishing	Yes									Yes	
3. Convention on high seas oil casualties								Yes		Yes	
4. RAMSAR	Yes	Yes	Yes	Yes		Yes				Yes	Yes
5. Dumping convention						Yes		Yes		Yes	Yes
6.Heritage Convention			Yes	Yes						Yes	Yes
7. CITES	Yes	Yes[a]				Yes				Yes	
8. Marine Pollution (not oil)								Yes		Yes	
9. Marpol						Yes		Yes		Yes	
10. CMS	Yes	Yes	Yes	Yes		Yes				Yes	
11. UNCLOS	Yes	Yes	Yes			Yes	Yes	Yes	Yes	Yes	Yes
12. Basel						Yes		Yes		Yes	Yes
13. CBD	Yes	Yes	Yes	Yes		Yes	Yes		Yes	Yes	Yes
14. High seas fisheries compliance	Yes									Yes	Yes
15. Part XI UNCLOS							Yes			Yes	Yes
16. Straddling stocks agreement	Yes	Yes			Yes	Yes		Yes		Yes	Yes
17. Protocol marine pollution					Yes	Yes	Yes	Yes		Yes	Yes
18. Cartegena							Yes		Yes	Yes	Yes
19. Stockholm						Yes	Yes	Yes		Yes	Yes
20. Antifouling					Yes	Yes	Yes	Yes		Yes	

Continued

Table 2 Characteristics of the international conventions and agreements that have a bearing on protection of marine biodiversity or the marine environment.—cont'd

Convention/agreement	A	B	C	D	E	F	G	H	I	J	K
21. Ballast						Yes	Yes		Yes	Yes	Yes
22. Port State measures	Yes									Yes	Yes
23. Nagoya	Yes	Yes	Yes							Yes	Yes

[a]Where trade in that species may impact on an endangered species.

Conventions and agreements are: (1) International Whaling Convention (1946); (2) Convention on Fishing and Conservation of the Living Resources of the High Seas (1958); (3) International Convention relating to Intervention on the High Seas in Cases of Oil Pollution Casualties (1969) (4) Convention on Wetlands of International Importance especially as Waterfowl Habitat (RAMSAR; 1971); (5) Convention on the Prevention of Marine Pollution by Dumping of Wastes and Other Matter (1972); (6) Convention Concerning the Protection of the World Cultural and Natural Heritage (1972); (7) Convention on International Trade in Endangered Species of Wild Flora and Fauna (CITES; 1973); (8) Protocol relating to Intervention on the High Seas in Cases of Marine Pollution by Substances other than Oil (1973); (9) Protocol of 1978 Relating to the International Convention for the Prevention of Pollution From Ships, 1973, (MARPOL); (10) Convention on the Conservation of Migratory Species of Wild Animals (CMS or Bonn Convention; 1979); (11) UN Convention on the Law of the Sea (UNCLOS; 1982); (12) Basel Convention on the Control of Transboundary Movements of Hazardous Wastes and Their Disposal (1989); (13) Convention on Biological Diversity (CBD; 1992); (14) Agreement to Promote Compliance With International Conservation and Management Measures by Fishing Vessels on the High Seas (1993); (15) Agreement relating to the implementation of Part XI of the United Nations Convention on the Law of the Sea of 10 December 1982 (1994); (16) Agreement for the Implementation of the Provisions of The United Nations Convention on the Law of The Sea of 10 December 1982 Relating to the Conservation and Management of Straddling Fish Stocks and Highly Migratory Fish Stocks (1995); (17) 1996 Protocol to the Convention on the Prevention of Marine Pollution by Dumping of Wastes and Other Matter, 1972 (1996); (18) Cartegena Protocol on Biosafety to the Convention on Biological Diversity (2000); (19) Stockholm Convention On Persistent Organic Pollutants (POPs) (2001); (20) International Convention on the Control of Harmful Anti-Fouling Systems on Ships (2001); (21) International Convention for the Control and Management of Ships' Ballast Water and Sediments (2004); (22) Agreement On Port State Measures To Prevent, Deter And Eliminate Illegal, Unreported And Unregulated Fishing (2009); (23) Nagoya Protocol on Access to Genetic Resources and the Fair and Equitable Sharing of Benefits Arising From Their Utilization to the Convention on Biodiversity (2010).

A. Sustainable management of living resources

B. Sustainable management of unexploited species

C. Habitat management or protection

D. Implement protected areas

E. Precautionary principle

F. Monitoring of species, habitats or environment

G. Environmental impact assessment

H. Prevention of environmental pollution

I. Biosecurity

J. Encourage or impel international cooperation

K. Capacity building

the distribution of marine species in many countries and in areas beyond national jurisdiction (e.g. Miloslavich et al., 2016; Wafar et al., 2011). This gap underlies the current calls for a moratorium on the initiation of mining in areas beyond national jurisdiction which prevents the development of meaningful environmental impact assessments prior to mining, but also makes monitoring the impacts of such activities after they have commenced extremely difficult and expensive (e.g. Amon et al., 2022). It is further inhibited by the lack of a biodiversity treaty that provides clear obligations to conserve marine biological diversity in areas beyond national jurisdiction. Even in the coastal waters of many nations, especially low- and middle-income countries monitoring changes in biodiversity is extremely difficult although there are exceptions at habitat level (e.g. the Global Coral Reef Monitoring Network [GCRMN] https://gcrmn.net/). Global indices used to track biodiversity, such as the IUCN Red List and the Living Planet Index suffer from lack of data and highly biased sampling of marine biodiversity (see above). In the following section of this paper we explore why our ability to discover marine species and to monitor biodiversity has progressed very slowly.

3. How marine biodiversity is studied today

3.1 Biodiversity and biodiversity science

The term 'biodiversity' originated as a derivative of the words 'biological diversity' almost three decades ago by the famous Harvard Zoologist EO Wilson (Wilson, 1985, 1988). It is now one of the most cited terms and important concepts in ecological research, environmental management and conservation (Cochrane et al., 2016). The Convention on Biological Diversity (1992) defines biodiversity as:

> The variability among living organisms from all sources including, inter alia, terrestrial, marine and other aquatic ecosystems and the ecological complexes of which they are part; this includes diversity within species, between species and of ecosystems.

Thus, marine biodiversity integrates ecosystem components, encompassing all levels of biological organization from genes and species to populations and ecosystems, with the diversity of each level having structural and functional attributes (Cochrane et al., 2016).

Defining marine biodiversity science is not straightforward as a variety of scientific disciplines examine biodiversity with very different perspectives and objectives. Cochrane et al. (2016) have helpfully summarised what biodiversity science comprises, and based on this we define it as:

> Biodiversity science refers to descriptions or assessments of the distribution, status and condition of all or selected groups of organisms, from their genetic variability, to the species, populations, communities, and ecosystems. It also encompasses understanding the interactions and functions on all levels from individuals up to the whole ecosystem, including changes related to natural and anthropogenic environmental pressures *(based on Cochrane et al., 2016).*

This view of biodiversity science is very much one coming from the science community; however, it neglects the role of nature in securing people's well-being (Bawa et al., 2020). Much of the Earth's biodiversity lies in tropical regions and within the waters of low- and middle-income countries (see above), and these are typically the regions that have historically been under resourced and suffered colonist practices. Without significant effort to build capacity in these countries to undertake biodiversity science and to assess the links between biodiversity and increased resilience to climate change and environmental shocks such as flooding, increasing human well-being and fostering socioeconomic development, halting biodiversity loss and conserving the natural environment will remain difficult (Bawa et al., 2020). There are also concerns that climate change impacts on marine ecosystems over the coming century will disproportionately impact biodiversity, and hence the social and economic benefits obtained. Low- and middle-income countries, with a higher dependence on such resources and a reduced capacity to respond and adapt will be disproportionately affected (Boyce et al., 2020, 2022). Moreover, awareness is not enough, as communities that depend on extractive practices for their subsistence are often faced with the dilemma of extract or conserve with no compensation. Biodiversity, more specifically, the blue natural capital that supports the benefits we receive from nature, should have a value whilst still in the ocean, not only when extracted, that improves the well-being and livelihoods of its custodians, to replace an extractive by a regenerative interaction with nature. The struggle to incorporate nature's value to people in considerations of how to halt biodiversity loss has also led to significant conflict within the Intergovernmental Science-Policy Platform on Biodiversity and Ecosystem Services (IPBES; Masood, 2018). It is therefore important that biodiversity science incorporates elements of social sciences, including equity, justice, and the concerns of Indigenous and local peoples, as well as economics to give a full picture of biodiversity and its interactions with human society. Widely accessible data and tools that enable stakeholders such as policy makers, industry and the public to make evidence-based decisions that benefit biodiversity are also important.

3.2 Taxonomy

3.2.1 The history of taxonomy

Underpinning many aspects of marine biodiversity research is the science of taxonomy. The classification of organisms probably dates back to prehuman times (Raven et al., 1971). The ability to classify plants and animals useful for food or medicine and those that may be poisonous or otherwise dangerous, and to pass this knowledge from generation to generation, has obvious survival value and may have played a role in human evolution. As human civilisations evolved across the world, folk taxonomies were developed to identify the organisms that were relevant to day-to-day life (Raven et al., 1971; Winston, 2018). Studies of folk taxonomies have shown that they have commonalities regardless of cultural and geographic origins such as the tendency to recognise naturally occurring groupings (e.g. plant, animal) and the subdivision of these into a range of hierarchical categories (Raven et al., 1971). These folk taxonomies would have formed the basis for the first attempts to record via the written word, knowledge on the classification of living things (Raven et al., 1971; Winston, 2018) and continue to be useful today (e.g. Medeiros et al., 2022).

Seminal works from all around the ancient world, including China, India and Europe shaped philosophical and scientific thinking around taxonomy through to the Renaissance (Manktelow, 2010; Winston, 2018). In Europe Aristotle (384–322BC) and his student Theophrastus (370–285BC) are credited with the first classifications of living things. The former was credited with defining vertebrates (animals with blood) and invertebrates (animals without blood) which reflected his approach of classifying organisms according to the similarity of their anatomical and functional biology (Richards, 2015). Theophrastus focused on the classification of plants (*De Historia Plantarum*) which contained about 480 species (Manktelow, 2010). Subsequent natural philosophers included the Greek Dioscorides (40–90AD) and the Roman Pliny the Elder (23–79AD). Dioscorides was a physician who travelled widely in Greece and Rome gathering information on medicinal plants, finally producing the work *De Materia Medica*, which was in use in medicine until the 16th Century (Manktelow, 2010). Pliny the Elder was a member of the Roman aristocracy who, following a career in the military, as a Governor and an advisor to the Roman Emperor, compiled a multi-volume work known as *Naturalis Historiae*. The work cited many previous Greek and Roman natural scientists and covers a broad range of the sciences including astronomy, natural history, geography and geology. Volumes 7–11 discussed zoology and in 12–19 botany (Rogers, 2019). The book was widely used as a basis for education on the natural world until the 15th Century.

Much of the work in Europe on classification of living organisms in medieval times through to the Renaissance was focused on plants, particularly herbs and medicinal plants. At this time much of the work on animals comprised bestiaries which included both real animals but also mythological beasts and were often used to illustrate religious morals, dogmas or values (Richards, 2015). The herbalists, on the other hand, were collecting together descriptions of species used by society (Richards, 2015). With the advent of printing and the fact that herbals sold well, there was a proliferation of publications on plant identification (Manktelow, 2010; Rogers, 2019). As the numbers of described plants grew, natural philosophers began to consider how to organise their classifications (Richards, 2015). Herbalists like Hieronymous Bock and Leonhart Fuchs tended to group similar plants together and the Bauhin brothers organised plants into genera containing species (Richards, 2015). Probably the most important figure during this period was Andrea Cesalpino (1519–1603) who developed a classification system based on similarities in growth habit and reproductive parts of plants (i.e. an Aristotleian approach of functionality), which was hierarchical although it only dealt with Genera and Species (Manktelow, 2010; Richards, 2015).

It is therefore apparent that by the time of the Age of Enlightenment some of the important foundations of Linnean Taxonomy were already in place. Some key figures at this time included Robert Morison (1620–1683) from Scotland, the German Augustus Rivenus (1652–1723), the English naturalist John Ray (1627–1705) and Joseph Pitton de Tournefort (1656–1708) from France. Morison, who became the first Professor of Botany at the University of Oxford, devised a system for the classification of plants emphasising the use of the structure of the fruits as well as publishing the first monograph on a specific group of plants: the Umbelliferae. Rivinus used the structure of flowers for classifications and used dichotomous keys for identification including of a higher taxonomic level of Order. John Ray produced several major works on taxonomy, including *Methodus Plantarum Nova*, which contained around 18,000 plant species (Manktelow, 2010). However, he is most notable as being the first person to associate the term 'species' with reproductive isolation and genealogy, not the possession of specific features (Richards, 2015). De Tournefort published *Institutiones Rei Herbariae*, in which around 9,000 species were listed in 698 genera (Manktelow, 2010). He emphasised the rank of Genus and many of his genera were subsequently used by Linnaeus and are still valid today (Manktelow, 2010).

At this time, global exploration, in search of lands for colonial invasions, was fully in progress and specimens of animals and plants were arriving on the shores of Europe from all over the world. This generated what has been termed the first taxonomic crisis (Godfray, 2002, 2007). With increasing numbers of species being described and scientists using differing approaches to classification, the growing science of taxonomy became chaotic. Fortunately, a young Swedish natural philosopher arrived on the scene with a passion for identifying, describing and classifying plants, Carl Linnaeus. An important aspect of his work was that he pressed flowers onto loose pages of paper which enabled him to organise and reorganise his dried specimens as he collected more samples and classified and reclassified them (Yoon, 2009). In 1735 at the age of 28 he published *Systema Naturae* (The System of Nature), a relatively short work (just 14 pages) that established the three Kingdoms of nature, plants, animals and minerals as well as Classes, Orders, Genera and Species (Richards, 2015; Yoon, 2009). This was followed by a series of regional studies on botany before *Species Plantarum* (1753) laid out a standardized approach to the nomenclature of plants throughout the world (Rogers, 2019). Through these works he established the binomial system for naming of organisms (Genus and Species name) and a system of rules for biological nomenclature (Rogers, 2019; Yoon, 2009). Linnaeus based his classification system on the reproductive structures of plants, an Aristotelian approach to classification, but in his writings indicated that this was preliminary and that a broader picture of the functional characters of species was needed for an accurate classification (Richards, 2015). He was also acutely aware that he had only examined a fraction of nature and so did not have a complete picture of the variation in species and higher taxonomic categories and thus his classification was likely to be incomplete (Richards, 2015). It is also notable that Linnaeus at first believed that 'species' were an immutable classification but later considered that perhaps new species could be created particularly through hybridisation of existing species (Richards, 2015). Linnaeus published new versions of *Systema Naturae* throughout his life, describing 7,700 species of plants and 4,400 species of animals by the time he completed his work (Yoon, 2009).

Linnaeus's work was not universally accepted at first. Particularly in France it was heavily criticised by the pre-eminent natural philosophers of the time. Georges-Louise Leclerc de Buffon (1707–1788) was a particular critic. Buffon believed that it was impossible to impose a hierarchical order on the disorderly natural world and that in reality the natural gradations between such categories were so subtle that they could not be defined

(Richards, 2015). Despite this, he did believe in a form of the biological species concept (Richards, 2015). Linnaeus's work was also not immediately accepted in Britain where the work of John Ray was still favoured, although later Linnaeus's work was accepted by the Royal Society (Rogers, 2019). By the second half of the 18th Century, Linnaeus's classification system was in wide use by British natural philosophers (Rogers, 2019).

Despite the advances in classification developed by Carl Linnaeus, the subsequent evolution of taxonomy was complex and not straightforward. Although there were advances on Linnaeus's classifications (e.g. Antoine Laurent de Jussieu [1748–1836] who developed a classification of plants based on a wider range of characters; Manktelow, 2010), natural scientists in different countries were giving the same species different names, or correcting the Latin of a previously described species giving rise to a plethora of synonyms (Godfray, 2002; Winston, 2018). Set against the background of colonialism, the 19th Century was the advent of exploration of the ocean using modern scientific approaches including mapping, charting, naming and collecting biological specimens all over the world. These expeditions included famous exploration ships such as the *HMS Beagle*, *Belgica*, *HMS Challenger*, *USS Enterprise*, the *Travailleur* and the *Valdivia* (Murray and Hjort, 1912; Rozwadowski, 2005). Together they provided natural scientists with a vast new treasure trove of biological specimens not only from all around the world but also from the depths of the ocean and from polar latitudes. Also, especially in Britain, a craze for middle-class amateur natural history focused on collecting on shores and dredging for marine life (Rozwadowski, 2005). Important marine biological stations were also set up across Europe, examples including the laboratories at Concarneau and Roscoff in France, the Stazione Zoologica in Naples, Italy, the Marine Biological Association laboratory, Plymouth, UK, and the Oceanographic Museum in Monaco, as well as in the USA (Marine Biological Laboratory, Woods Hole) and by the end of the 19th century, in Japan (Misaki, Kanagawa Prefecture; Inaba, 2015). This was also the time of the establishment of several of the most significant institutions globally for housing collections of specimens of organisms and for taxonomic research including the Natural History Museum of London, the Muséum National d'Histoire Naturelle in Paris (established at the end of the 18th Century), the Smithsonian Institution (later opening the National Museum of Natural History) and the Raffles Museum in Singapore (later the National Museum of Singapore). The legacy of this period is a plethora of synonyms (multiple names for the same species), inadequate descriptions for marine species or even worse, species names with no descriptions (*nomen nudum*) as well as descriptions with no deposited type material. This legacy is a contributor to the current taxonomic crisis (see below; Godfray, 2002).

Before the end of the 19th century, it was recognised that efforts were required to systematise the rules for the naming and classification of species. Zoological and botanical naming practices developed along different paths. The Strickland Code (Strickland, 1843) for the nomenclature of animals developed first (Winston, 2018). This was adopted in Britain, the USA and Italy but not elsewhere. The first botanical code was developed by the Swiss botanist Alphonse de Candolle in 1867 but, again, this was not adopted universally. Following the First International Congress on Zoology in Paris in 1889, it was agreed to develop an international zoological code (Winston, 2018). The first version of the International Code for Zoological Nomenclature was published in 1905 with the International Code for Botanical Nomenclature following in 1906 (Winston, 2018). These International codes have been updated several times since their initial versions were published and are managed through commissions (e.g. the International Commission on Zoological Nomenclature (ICZN) whose secretariat is now based at the Lee Kong Chian Natural History Museum in Singapore; Winston, 2018). These Commissions make decisions on cases related to taxonomy where scientists encounter problems in nomenclature (e.g. the priority name for a species is the oldest one and sometimes this is not clear; Winston, 2018). Microbiologists working on prokaryotes established a naming system based on the botanical nomenclature system much later in the 20th century although because of the complications related to past work on microbial taxonomy they made a decision to invalidate all names previous to 1980 (Winston, 2018). A separate nomenclature system has subsequently been developed for viruses (Winston, 2018). In order to try and bring together the various codes of nomenclature the International Union of Biological Sciences (IUBS) has tried to establish the Biocode which is intended to apply to all organisms other than viruses (Winston, 2018). The latest version is the 2011 Biocode (Greuter et al., 2011).

More recently, because of a lack of agreement from cladistic taxonomists on the use of the rank-based nomenclature of Linnaeus there has also been an attempt to establish a new code, the Phylocode (De Queiroz and Gauthier, 1990, 1994). Phylocode is based on cladistics, an approach to biological classification based on grouping organisms according to most common recent ancestry, the evidence for which is in the form of shared characteristics (shared derived characters or synapomorphies). The method was an attempt to move away from a system of taxonomy where scientists applied their knowledge of functional characters of organisms to decide which formed natural groupings. The latter approach was viewed as subjective, especially in terms of deciding which characters were important. Also, proponents of Phylocode point to the artificiality of establishing a rigid hierarchy of classification, a product of 18th Century religion and philosophy, rather than a system that more closely reconstructs the variation of relationships between organisms which have arisen through evolution (Pennisi, 2001). Cladistics generates cladograms, essentially a phylogenetic tree representing the best hypothesis of phylogenetic relationships. However, in its early stages such trees were based on models of parsimony which minimised the number of evolutionary changes to produce the cladogram. Now maximum likelihood models are viewed as a more accurate representation of evolution. This is especially the case with the advent of widely available and cheap DNA sequencing data for testing the relationships between species. Phylocode has been extremely controversial (e.g. Pennisi, 2001) and rejected by many taxonomists (e.g. Nixon et al., 2003). However, the latest version of Phylocode has been published (Cantino and de Queiroz, 2020) and as yet its role in taxonomy is not clear.

3.2.2 Taxonomy in the digital age

The late 20th Century brought a further wave of change in taxonomy. This is through the digitization of taxonomic information and the development of the internet. Firstly, many paper-based scientific journals and monographs, previously only printed on paper, came to be available in electronic form, mainly as pdf files (Portable Document Format). In an increasing number of cases, journals are entirely electronic, or they are published in both electronic format and as paper-printed versions. In 2012, the ICZN allowed publication of new names and nomenclatural acts in widely accessible electronic media provided that these were registered in ZooBank, a new database (accessible at: http://zoobank.org/; Winston, 2018). The International Code of Nomenclature for Algae, Fungi, and Plants followed in 2018 (Winston, 2018). ZooBank is also now working on a List of Available Names (LAN), with the aim of providing a definitive list of names for taxa with a structure of quality assurance codes (Winston, 2018).

As well as the attempts to bring the commissions/codes for nomenclature into the digital age, there has been an enormous effort in cataloguing biodiversity data on the web. These efforts vary from global projects aimed at cataloguing all species, occurrence records and biodiversity datasets (e.g. Global Biodiversity Information Facility or GBIF, https://www.gbif.org/; Encyclopaedia of Life or EOL, https://eol.org/), digital sequence information (e.g. the International Nucleotide Sequence Database Collaboration, INSDC, http://www.insdc.org/), to efforts focused on particular biomes (e.g. Ocean Biodiversity Information System, https://obis.org/) or groups of organisms (e.g. FishBase, http://wpbp.sinica.edu.tw/search.php). Some of these databases hold very specific types of data (e.g. INSDC, Fishbase), whilst others hold a multiplicity of data types (e.g. GBIF holds institutional metadata sets, checklists of species names, species occurrence data and quantitative sampling event data). Such a multiplicity of data sources constructed for various uses, mainly for scientists but also for other purposes such as education or citizen science (e.g. i-Naturalist, https://www.inaturalist.org/), have introduced further confusion in terms of where and how to access biodiversity data. Inevitably there has been competition for resources and some of these tools have collapsed. An example is the Tree of Life Web Project (TOLWEB) which was aimed at providing information on the diversity and phylogeny of life on Earth. Issues occurred with the validation of trees and other information on the site by voluntary experts and eventually the project simply ran out of funding.

One way to counter the issues with the growing number of biodiversity databases is to make them interoperable and to try and reconcile records across multiple platforms. An example of this has been the Catalogue of Life Plus project (COL). This project aims to resolve the issue whereby large database projects (e.g. GBIF and EOL) catalogue species names to achieve their own objectives but in the process introduce inconsistencies in nomenclature (COL, 2018). The project aims to create an open, shared, and sustainable consensus taxonomy to serve the proper linking of data in the global biodiversity information initiatives, effectively a clearing house for taxonomy and nomenclature (COL, 2018). Text mining is a growing field in Artificial Intelligence, which is starting to find applications in biodiversity science, mostly to extract data from the published literature (Cornford et al., 2021; Heberling et al., 2019; Le Guillarme and Thuiller, 2022), and has the potential to help assimilate information from a diverse range of databases and other sources.

3.3 The future of taxonomic research

The end of the 20th Century found biodiversity research facing what has been called a second taxonomic crisis (Godfray, 2002). The success of academics and academic institutions in the developed world are now largely judged by proxies for the quality of science produced, such as measures of journal impact factors and citations (Pinto et al., 2021), funding accrued for projects and, for universities, teaching quality. Unfortunately, as an arms race developed between peer-reviewed scientific journals to acquire the highest impact factors, taxonomic work was downgraded because of its very specialised nature and low numbers of citations (Pinto et al., 2021). As a result, it became increasingly relegated to specialised journals with low impact factors, and some have been threatened with de-listing from journal citation indexes because of high levels of self-citation (Hamilton et al., 2021; Pinto et al., 2021). At the same time, biological sciences were undergoing a significant expansion with the molecular revolution affecting many branches of science, particularly with the advent of cheaper and faster DNA/RNA sequencing. Taxonomists, many working on their own specific branch of the tree of life, often struggling with the legacy of 250 years of synonyms, inadequate descriptions and a lack of type material, were slow at producing work and not perceived as addressing big scientific questions of relevance to society (Godfray, 2002, 2007; Godfray and Knapp, 2004). As a result of this funding and supporting infrastructure for taxonomists and taxonomic research was greatly decreased (Baldini et al., 2021; Godfray, 2002, 2007; Saunders, 2020). In addition, many governments have also reduced direct funding to national museums, major centres of taxonomy, not only reducing opportunities for taxonomic work but, in some cases, endangering or indirectly causing the loss of globally valuable collections of specimens (Dupérré, 2020). Furthermore, aspiring young scientists were also put off studying taxonomy because of its poor career prospects. Added to this were the increased permitting and reporting requirements brought about by the Convention for the International Trade in Endangered Species and the Nagoya Treaty which was established to control access to and benefit sharing from genetic resources following reports of abuses. These treaties introduced permitting processes which have, for various reasons stifled specific areas of marine taxonomic research (e.g. on tropical reef-forming corals) or work in regions where governments have deliberately prevented permitting or do not have the capacity to administer it (e.g. Britz et al., 2020; Kursar, 2011).

For biodiversity research, including in the ocean, the second taxonomic crisis has undermined our ability to understand the current crisis of biodiversity loss (see above). A lack of taxonomists and funding for taxonomy means that understanding biodiversity baselines in many parts of the ocean, as well as trends of biodiversity loss, is immensely challenging. We cannot hope to understand the big picture for marine life without a more comprehensive footing in terms of knowledge about the biodiversity in the ocean—how much there is, where it is, and how it is at risk. The old approach of slowly working through descriptions of animals is simply not up to this challenge, because of the issues outlined above. With only ~10% of marine species described and many parts of the ocean still relatively unexplored and un-investigated in terms of marine biodiversity, especially in the southern hemisphere, the coastal waters of developing countries and in the deep ocean, the challenge is truly enormous. Responding to this challenge, however, it is important to avoid the mistake of proposing projects which are simply unachievable (Godfray, 2002), and also important to clearly communicate to funders, policymakers and the public why this work is important (Saunders, 2020). Hence new approaches are needed.

Conventionally, marine biodiversity research is undertaken through sampling organisms in a quantitative manner from the marine environment. Traditionally this has meant collecting actual organisms, killing and preserving them and then returning them to the laboratory for identification and, if necessary and possible, description of new species (Rogers et al., 2021). For examples of sampling methods and considerations see Selvanes et al. (2017), for plankton see Harris et al. (2000) and for the deep sea see Clark et al. (2016) and Woodall et al. (2018). The processes of species identification and description require specialist knowledge of external or internal morphological characters to compare individual animals with those which have already been described, either assigning them to known species or writing a detailed description of new species for publication in a scientific journal. Such scientific descriptions can be lengthy and are written more or less in a formulaic way, often with photographs and/or diagrams of defining features. Sometimes notes can be associated with a species description on the habitat in which the organism is found and its ecology. When the majority of species in samples from some parts of the world or some environments are undescribed (e.g. the deep sea) such work can take a long time; often many years for a single group of organisms from a single expedition. This approach will not solve the crisis in terms of evaluating the loss of biodiversity from the ocean, especially with the current lack of resources and human capacity for taxonomy. Newer technologies and approaches which can greatly assist species identification, or identification to higher taxonomic levels (i.e. genus, family) fall into a number of categories:

- Imagery technologies
  - Photography and videography
  - Plankton counters
  - Micro-CT and other 3D-imaging approaches

- Molecular approaches
  - Biochemical genetics
  - Molecular barcoding
  - Environmental DNA (eDNA)
  - Genome sequencing and metagenomics
- Acoustic approaches
  - Multibeam and multibeam backscatter data
  - Active biological acoustics
  - Passive acoustics
- Databases
- Citizen science

3.3.1 Imagery technologies

At their most basic, surveys of marine ecosystems to assess biodiversity or other aspects of community ecology can use Underwater Visual Census (UVC). This method is usually used by SCUBA or technical divers to visually identify and estimate the abundance of fish populations underwater either along a transect or by static count (e.g. Bernard et al., 2013; Prato et al., 2017; Samoilys and Carlos, 2000). The method is subject to various biases related to the behaviour of specific types of fish, the exact methods deployed and also the skills and experience of the people undertaking the census (e.g. Prato et al., 2017; Samoilys and Carlos, 2000). There are also technical limitations in terms of depth and other aspects. However, UVC has significant advantages in its ability to gather large quantities of data, quickly, with relatively little equipment. Standardized UVC approaches from numerous sites have been combined to unveil large-scale patterns in biodiversity (Stuart-Smith et al., 2013). UVC has been used relatively little for survey of benthic biodiversity although it has been used to estimate habitat complexity and benthic cover by sessile organisms (e.g. Wilson et al., 2007a).

Most other non-extractive forms of survey for biodiversity research use photography or video-based survey methods. These have been deployed from the second half of the 20th Century for a wide range of purposes but in the last two decades have increased with the dramatic improvement in technology including in underwater housings, image quality, battery longevity, reduced cost, and data storage (Bicknell et al., 2016). They are now deployed routinely for many ecological studies for example in ecosystem and population monitoring, assessment of communities in marine protected areas and fisheries management and enforcement, (Bicknell et al., 2016).

Cameras have been deployed in many forms, but common technologies have included the use of photographs or video for surveying of benthic habitats (e.g. photoquadrats, drop-cameras and photo and video transects from an array of platforms; Cleguer et al., 2021; Davies et al., 2015; Edmunds et al., 1998; Hartwell et al., 2018; Isaacs and Schwartzlose, 1975; Laverick et al., 2017; Morris et al., 2012; Parravicini et al., 2009; Ramirez-Llodra et al., 2020; Rice et al., 1979; Schofield et al., 2019; Stefanoudis et al., 2019a) or for surveying of fish or pelagic communities (e.g. Baited Remote Underwater Video; video systems deployed from a variety of platforms; Andradi-Brown et al., 2016, Forrest et al., 2020; Hoving et al., 2019; Stefanoudis et al., 2019b; Sward et al., 2019).

Collection of photographic and video data using modern, high-resolution cameras requires large data storage capacity and the need for extracting data from the images (Bicknell et al., 2016). A single expedition can generate potentially hundreds of hours of geo-referenced video data. In the past these data were then extracted manually by a scientist slowly working through the video and recording when organisms are encountered (Bicknell et al., 2016). This is obviously an extremely time-demanding procedure, and as more image data are collected it is becoming less practical. Currently an approach which is showing some success in the identification of organisms from image data is Machine Learning, a subfield of Artificial Intelligence (Beyan and Browman, 2020). In particular, a branch of machine learning, Deep Learning, is being applied to identification of marine species through the use of Artificial Neural Networks (Beyan and Browman, 2020). This has been enabled by the increase in computing power offered by massively parallel processing units but also by the availability of new computer algorithms that can use these computer architectures and the availability of large training datasets (Beyan and Browman, 2020; Wäldchen and Mäder, 2018). Artificial Neural Networks require training on images (i.e. training datasets) to allow them to develop dynamic models to classify image data (Inference; Beyan and Browman, 2020; Villon et al., 2018; Wäldchen and Mäder, 2018).

Automated classification of fish and other types of organisms faces many challenges such as variation in lighting conditions, distortions like poor image quality and noise, changes in orientation and size of the object of interest in the image and variations in the background, all of which impose non-linearity in the image data distribution (Salman et al., 2016, 2019). Such approaches generally require two distinct steps: (i) Feature extraction (e.g. discriminating fish from non-fish objects) (ii) Classification (i.e. taxonomic identification; Salman et al., 2016, 2019; Wäldchen and Mäder, 2018). Sometimes other tasks can be undertaken following classification such as measuring fish length and estimating biomass (Salman et al., 2016, 2019). Deep Learning approaches using Neural Networks have been increasingly applied to the identification of a variety of marine organisms including phytoplankton (Li et al., 2020), zooplankton (Campbell et al., 2020) and corals (Gómez-Ríos et al., 2019). This has included identification of organisms from 3-dimensional images generated by structure-from-motion photogrammetry (e.g. for coral reefs; Hopkinson et al., 2020). For some marine ecosystems, fully or semi-autonomous extraction of images and identification of organisms is now routinely used for some taxa (e.g. coral using CoralNet or Squidle+). Neural Network approaches also lie behind a number of digital identification tools available as mobile phone Apps. Examples include: i-Naturalist (https://www.inaturalist.org/), PictureThis (https://www.picturethisai.com/) and PictureBird (https://picturebirdai.com/). Whilst the latter two Apps are available as free tools in basic format they also include In-App purchases to enhance performance (e.g. to provide gardening information). i-Naturalist is free and is a community-based tool for wildlife identification. A discussion of Deep Learning approaches to the classification of organisms is presented in Wäldchen and Mäder (2018).

3.3.2 Plankton counters

This includes a range of towed, autonomous and lab-based technologies for counting and in some cases identifying plankton. At their simplest, these devices comprise of a light source (e.g. a Light Emitting Diode; LED) and a light detector (photodiode) and are a type of Optical Plankton Counter. As plankton pass in front of the detector the light is attenuated and this signal indicates the number of plankton and their length (Herman, 1988, 1992). Later versions of this device employed a laser as a light source and could produce a cross section image of passing particles (plankton) on a photodiode (Herman et al., 2004). Increasing resolution allowed for increased ability to discriminate zooplankton taxa. These early types of Optical Plankton Counters were mounted on towed platforms (e.g. batfish) and could undertake quantitative profiles of zooplankton in the ocean. In the late 1990s, Video Plankton Recorders and another imaging device, the Shadowed Image Particle Profiling and Evaluation Recorder were developed (Basedow et al., 2013; Broughton and Lough, 2006). These provide information on plankton abundance but also on the taxonomic identification of plankton communities (Broughton and Lough, 2006). Both Optical Plankton

Counters and Video Plankton Recorders can suffer from inaccuracies as a result of their design with plankton orientation being an issue with the former and focus an issue with the latter; both also only sample a small volume of water (Basedow et al., 2013). Various iterations of Video Plankton Recorders have been developed (see Lombard et al., 2019 for review) and Deep Learning approaches to data analysis (species identification) applied (e.g. Davis et al., 2005). A new development in this field of technology is the use of holographic cameras for the imaging of phytoplankton, zooplankton and other particles in the ocean such as microplastics (see Nayak et al., 2021).

An alternative approach to identification of planktonic organisms has been to sample them using conventional methods and then to use image analysis to identify species in the laboratory. These technologies essentially involve producing an image of the organisms which is then analysed using software algorithms to identify species (e.g. Zooscan; Gorsky et al., 2010). FlowCAM combines flow cytometry with microscopy to count and size small particles but also to produce images of them (e.g. Álvarez et al., 2014; Sieracki et al., 1998). Potentially tens of thousands of images can be collected using this technology with scientists visually identifying the dominant groups of organisms (Jakobsen and Carstensen, 2011). Up until recently, automatic identification of images has been quite limited in accuracy (e.g. Álvarez et al., 2014). This is being improved through use of collaborative Deep Learning approaches (e.g. Kerr et al., 2020). Flow Cytobot is another type of imaging flow cytometer that can be built into moorings, buoys or other types of platforms for in-situ measurements of plankton community dynamics (Olson and Sosik, 2007). Classification of plankton with this machine is carried out using a Support Vector Machine (SVM) approach, another type of Machine Learning algorithm (Sosik and Olson, 2007).

3.3.3 Micro-CT and other 3D-imaging approaches

One of the main bottlenecks in taxonomy is the availability of type material (Faulwetter et al., 2013a,b). Even if type material exists, it can be very difficult to obtain as a loan from a museum and assembling material for a large paper describing multiple new species or for a major taxonomic revision can take a great deal of time or be near impossible (Faulwetter et al., 2013a,b). One solution is to generate high quality three-dimensional (3D) images of a type specimen, a 'cybertype' (Godfray, 2007). Non-destructive three-dimensional imaging techniques such as Confocal Laser Scanning Microscopy, Optical Projection Tomography, Magnetic

Resonance Imaging and Micro-Computed Tomography (micro-CT), can rapidly create high-resolution morphological and anatomical data in 3D (Boistel et al., 2011; Faulwetter et al., 2013a,b). Whilst these methods have been criticised as having insufficient resolution for taxonomic purposes (e.g. Dupérré, 2020) and there is some evidence to support this in terms of the current resolution of technologies (Faulwetter et al., 2013a,b) many of these fields are advancing rapidly and resolution is improving. Other criticisms of the use of 3D digital images for taxonomy include that they do not represent the full range of characteristics required for diagnosis of a species and that they cannot replace type specimens (the examples of a species deposited at a museum or other repository upon which a species description is based; Dupérré, 2020; Faulwetter et al., 2013b; Gutiérrez and Pine, 2017). Faulwetter et al. (2013b) suggest a number of measures to ensure that 3D-images are appropriate as cybertypes including:

- A cybertype should provide the same information at the same level of resolution and accuracy as a conventional morphological type specimen for a species description. This includes all the external and internal anatomical features of the organism.
- A cybertype should be linked to the original type material which can be consulted if there is doubt with respect to comparisons on specific characters.
- A cybertype should be easily found, free to use and easily accessible (e.g. under a digital commons license and according to FAIR principles; Findable, Accessible, Interoperable and Reusable; Wilkinson et al., 2016b).

For many marine invertebrates, for example, dissection of specimens (e.g. echiuran worms; Rogers and Nash, 1996) or fixation and sectioning of entire animals or parts of them (e.g. nemertean worms; Rogers et al., 1993) is required to reveal aspects of internal anatomy for a full taxonomic description. One significant advantage of techniques such as micro-CT is that they do not require destruction of a physical specimen (e.g. Faulwetter et al., 2013a,b). This retains the type material in an intact state for the future. Some methods such as Magnetic Resonance Imaging and Optical Projection Tomography are also possible using live specimens of organisms (Boistel et al., 2011).

3.3.4 *Molecular approaches*

3.3.4.1 Biochemical genetics

Linnean taxonomy developed prior to Darwin and an understanding of evolution. At this time classification was already based on the idea of the use of similarity to group organisms (see above). It was only with an understanding of evolution that scientists realised that groups of similar species were related by recent common ancestry. An issue with morphological taxonomy as practiced in the traditional sense was that the selection of characters on which to base similarity and to diagnose and describe new species was subjective. To some extent this was overcome by the invention of cladistics, which standardised approaches and mathematical procedures for the objective analysis of morphological characters to estimate similarity between taxa (Thorpe and Solé-Cava, 1994). However, cladistic approaches rely on the coding of morphological characters that are homologous; in other words, they have the same evolutionary origin, as well as issues of what characters are selected and the issue of parsimony (see above). An obvious way to understand the relationships between taxa is through genetic information. Until relatively recently, obtaining genetic information was restricted to the use of Pedigree Charts or through complex breeding experiments to obtain data on variation in phenotype (usually morphological variation). However, the development of isozyme electrophoresis through coupling of enzyme electrophoresis and histochemical staining (Hunter and Markert, 1957) provided a new method to estimate gene frequencies in natural populations (Thorpe and Solé-Cava, 1994).

Enzyme electrophoresis is based on the differential migration of water-soluble proteins (usually enzymes) through a medium (e.g. a gel of hydrolysed starch, cellulose acetate or polyacrylamide) under the influence of an electric potential. The migration of proteins is proportional to their charge-to-mass ratio, the properties of the gel and the potential difference applied. Mutations in genomic DNA which codes for a protein can cause a change in amino-acid sequence and this in turn affects the migration of the protein. Thus, in a protein which is heterozygous in a population (more than one variant as a result of mutation), the alleles can be detected as they have different mobilities. Most proteins or enzymes are codominant markers, in other words both variants in an individual can be detected in diploid organisms. Proteins are visualised through the specific histochemical stains so that a single enzyme can be visualised. About one in three mutations in genomic DNA leads to a change in amino acid sequence and not all of these will result in a change to the charge to mass ratio. Therefore, enzyme electrophoresis is an indirect method of assessing allele frequencies in a population.

Gene frequency data obtained from multiple enzymes can be used to estimate the genetic distance or genetic identity (similarity) between populations of organisms (Thorpe, 1982). It was discovered that because genes (and therefore gene products) exhibit clock-like behaviour in terms of mutations and changes in allele frequencies over time, levels of genetic distance and identity were relatively consistent with the taxonomic separation of populations (intraspecific, interspecific and intergeneric; Thorpe, 1982). Enzyme electrophoresis could therefore be used in many cases to provide a relatively objective guide to whether or not populations of organisms belonged to the same species or whether they represented different species likely in the same genus or in different genera (Thorpe, 1982). For genetic identity (*I*) about 98% of intraspecific *I*-values were found to lie above 0.85 whilst 97% of interspecific comparisons lie below 0.85 and for species in different genera about 77% fell below 0.35 (Thorpe, 1982; Fig. 7). This meant that there were some cases of comparisons between species where the genetic identity was not conclusive (i.e. conspecific and interspecific comparisons overlap to a small degree), a problem which is analogous to the lack of a barcoding gap in DNA barcoding for some taxa (see below). This reflects the natural processes of evolution where populations may be recently separated, hybridisation may take place or other genetic phenomena which alter the relationship between genetic identity (*I*) and taxonomic separation

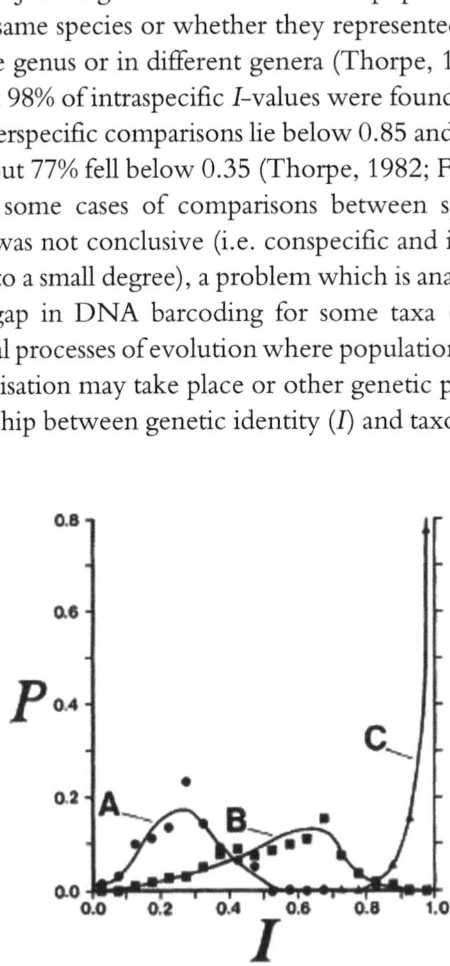

Fig. 7 Graph of probability vs Nei's (1972) Genetic Identity (I) for confamilial genera (A, black circles), congeneric species (B, black squares) and conspecific populations (C, black triangles). Note the considerable overlap in I for confamilial genera and for congeneric species and smaller zone of overlap for congeneric species and conspecific populations. *From Thorpe, J.P., 1982. The molecular clock hypothesis: biochemical evolution, genetic differentiation and systematics. Annu. Rev. Ecol. Syst. 13, 139–168.*

make using this method difficult (see Thorpe and Solé-Cava, 1994). There is a bigger overlap in comparisons between species within genera, and in different genera and beyond the taxonomic level of genus this type of marker becomes saturated (i.e. reverse mutations or mutations exhibiting the same phenotype become prevalent obscuring any useful evolutionary signal) and yields little useful information.

Enzyme electrophoresis has been widely used for molecular taxonomic studies in marine species. It is able to provide information at the intraspecific level, as to whether populations within a species are freely interbreeding or not (e.g. as a guide to whether populations of fish formed genetically distinct stocks for management purposes). It was also extremely useful in the identification of cryptic species (i.e. species which are distinct, but which show little morphological differentiation; see Thorpe and Solé-Cava, 1994 for examples). A further point related to enzyme electrophoresis is that enzymes are the phenotype of the genes which encode for them and are therefore subject to selection. This can interfere with their use as taxonomic guides as the molecular clock assumes neutral or nearly neutral evolution. However, it can also be useful in terms of studying the selective effects of environmental variation on speciation and adaptation (e.g. temperature selection of thermostable enzyme alleles in hydrothermal vent polychaetes; Jollivet et al., 1995; Piccino et al., 2004).

Whilst the biochemical genetic approach of enzyme electrophoresis led to the first wave of 'molecular barcoding' it does have disadvantages. Firstly, only a limited number of mutations in the DNA coding for enzymes are detectable, limiting the application of the method. Secondly, only a few genetic loci are screened meaning that only a small fraction of the genome is accounted for. As a result, enzyme electrophoresis often requires substantial sample sizes. Thirdly, enzyme electrophoresis also requires fresh or frozen material as proteins degrade rapidly once isolated from an organism. Extracting proteins for enzyme electrophoresis also requires taking a significant piece or biopsy of tissue from an individual to be sampled or, with smaller organisms the complete maceration and loss of the specimen. From a safety point of view, many of the chemicals used for histochemical staining are hazardous.

3.3.4.2 Molecular barcoding

Molecular barcoding is a microgenomic identification system for species based on using a small segment of the genome which is sequenced (Hebert et al., 2003a,b). Given that most species last for millions of years and assuming clock-like behaviour of the mutation process, then a number of changes would be expected in nucleotide sequence between species (Hebert et al., 2003a). The genetic code, based on four types of nucleotides, means that billions of combinations of unique sequences may exist even across a short stretch of DNA of several hundred base pairs in length across taxa (although in reality many nucleotide positions are constrained by selection; Hebert et al., 2003a). These unique sequences for individual species are known as DNA barcodes or molecular barcodes (Hebert et al., 2003a,b).

Molecular barcoding was first adopted by scientists working on microbial organisms because they are so difficult to identify using morphological characters (Hebert et al., 2003a). However, as the second taxonomic crisis manifested, scientists recognised that it may provide a means to speed up diagnosis of species in larger organisms, especially when faced with the challenges of a large proportion of undescribed biodiversity and the issue of cryptic species (Hebert et al., 2003a). Very rapidly, scientists focused on the mitochondrial genome as being highly suitable for molecular barcoding. This was because of its lack of introns, its limited exposure to recombination, its haploid mode of inheritance (Hebert et al., 2003a) but also its relatively high mutation rate and the fact that it exists in multiple copies in every cell making sections of it relatively easy to amplify using the polymerase chain reaction (PCR) and then to sequence. The region of the mitochondrial genome coding for cytochrome oxidase I (COI) was identified as a particularly interesting region for molecular barcoding for most taxa. This is because it has highly conserved flanking regions enabling the design of primers that work for a wide range of taxa ('universal primers') and because the gene appeared to have a greater range of phylogenetic signal than other mitochondrial genes (Hebert et al., 2003a). Early tests of molecular barcoding using COI indicated that it had good discriminatory power at a range of taxonomic levels from phylum to species, although this study was focused mainly on insects (Hebert et al., 2003a). Initial analyses of Australian fish (Ward et al., 2005) and other taxa indicated that sequencing this region of the mitochondrial genome could resolve 95–97% of taxa (Hebert and Gregory, 2005). Within a very short time, a number of projects were established to gather barcodes for all representatives of specific taxa (e.g. Fish Barcode of Life project; FishBOL; Ivanova et al., 2007; Ward et al., 2009) and also the development of the Barcode of Life Data System (BOLD; http://www.boldsystems.org/index.php; see Ratnasingham and Hebert, 2007). The rapid spread of the idea was supported by a biannual international symposium on molecular barcoding commencing in 2005 (Adamowicz, 2015).

Despite the apparent efficacy of molecular barcoding, the approach was met with controversy. Criticisms were mainly aimed at the analyses of sequence data for identification of species or placement in higher taxa (e.g. DeSalle et al., 2005; Will and Rubinoff, 2004) or in the way DNA barcoding studies were designed and executed (e.g. Collins and Cruickshank, 2013). There were also concerns that DNA barcoding would supplant or undermine traditional taxonomy rather than being deployed as a useful tool for taxonomists (e.g. Ebach and Holdrege, 2005; Will et al., 2005). Some significant challenges in DNA barcoding have manifested, including the existence of a barcoding gap, taxa where COI shows low rates of molecular evolution and the existence of large numbers of barcodes in databases which are mislabelled/incorrectly attributed to taxa. These are described below.

Molecular barcoding depends on a barcoding gap, a clear gap between intraspecific genetic distances and interspecific genetic distances at least for the large majority of taxa (e.g. Meyer and Paulay, 2005). Initial work across taxa, as well as focused efforts on birds, spiders and fish indicated that the barcoding gap existed, and that COI was an effective region to delimit species (e.g. Barrett and Hebert, 2005; Hebert et al., 2003a,b, 2004; Ward et al., 2005) with the exception of the Cnidaria (see below). However, subsequent studies indicate that intraspecific variation and the prevalence of young species may have been significantly underestimated in these studies because of a lack of adequate sampling of both multiple individuals of a species as well as sampling from across geographic ranges (Meyer and Paulay, 2005). A detailed study of a well-sampled taxon, the cowries, along with trochid gastropods and limpets revealed that error rates could be substantial (Meyer and Paulay, 2005). In cowries the total error rate at the best threshold value for divergence was about 17%, whilst for trochids it was lower at 7% and for limpets all species could be resolved using COI (Meyer and Paulay, 2005). The differences in the groups were related to contrasting life histories which lead to differences in evolutionary processes between the taxa leading to higher intraspecific variation and the occurrence of ancestral polymorphisms in 'young' species of cowries (Meyer and Paulay, 2005). It is also important to point out that this figure was when defining species using a phylogenetic species concept (the smallest set of organisms that share a common ancestor and which can be distinguished from other groups; Evolutionary Significant Units or ESUs); if traditional concepts of cowrie taxonomy were applied, based on morphology this error rate doubled (Meyer and Paulay, 2005). Sources of error in cowrie assignment arose from imperfect taxonomy (e.g. non-recognition of species or from inappropriately assigning species to morphotypes; overlumping and oversplitting), or from incomplete lineage sorting where recently evolved species still share haplotypes (Meyer and Paulay, 2005). Other sources of error including the occurrence of nuclear copies of mitochondrial genes (NUMTs), inadequate sampling of taxa and introgression as a result of hybridisation were not significant sources of error in this study but may be in others (Meyer and Paulay, 2005).

For some taxa, most notably Cnidaria and especially the Anthozoa, but also for sponges, rates of molecular evolution are significantly lower than is usually the case (e.g. Hebert et al., 2003b; Huang et al., 2008; Mcfadden et al., 2011). In Cnidaria, this has been linked to a DNA-repair gene located in the mitochondrial genome (Hellberg, 2006; Pont-Kingdon et al., 1998) which was probably acquired early in the evolution of the phylum by horizontal gene transfer from a prokaryote (Bilewitch and Degnan, 2011). For octocorals, for example, levels of intraspecific variation at COI (as well as two other mitochondrial regions) were low and there was no clear barcoding gap between intraspecific and interspecific comparisons (Mcfadden et al., 2011). However, even though variation is low, species which diverge by more than 0.5% nucleotide sequence should be investigated as to whether they are different species as there is some taxonomic signal even with such low genetic variation (Mcfadden et al., 2011). In other cases, other mitochondrial genes than COI have proven to be more reliably amplified and provide better taxonomic resolution (e.g. hydroids, see Moura et al., 2008, 2011). However, part of the efficacy of COI as a molecular barcoding region is that this single microgenomic region can be an effective tool for species identification across many taxa (Hebert et al., 2003a). Applied as a rigorously standardized sequence of a minimum length and quality from an agreed-upon gene, deposited in a major sequence database, and attached to a voucher specimen whose origins and current status are recorded is an important advantage of focusing on a single microgenomic region (Hebert and Gregory, 2005).

A very serious issue with the current Barcode of Life Data System (BOLD) is the trustworthiness of the data and its curation. A recent hackathon of several large taxa in the BOLD (crustaceans, echinoderms, molluscs, and polychaetes) found that 39% of records were tagged as misidentified, mislabelled, contaminated, ambiguous, complex (species occurring with multiple barcode index numbers) or species with shared barcode sequences (Radulovici et al., 2021). It was suggested that such a high number of records which are problematic indicates flaws in the barcoding workflow (e.g. misidentifications or cross-contamination) or taxonomic issues (e.g. synonyms or unidentified species; Radulovici et al., 2021). The authors of this study pointed to the extremely large effort it required and that such resources could not be available on a long-term basis. It was suggested that automated approaches, using Machine Learning or Artificial Intelligence algorithms could greatly assist in automatically sifting out such problems, although some would still need human curation to resolve (Radulovici et al., 2021).

3.3.4.3 Environmental DNA (eDNA)

Environmental DNA (eDNA) is the complex mixture of DNA of many different organisms found in an environmental sample (e.g. seawater, sediment, faecal sample; Taberlet et al., 2018). eDNA is used to obtain information on the biodiversity of the environment from which it is sampled or on the functional diversity present (Taberlet et al., 2018). The latter is usually undertaken through metagenomic approaches and this will be covered in the next section. eDNA includes both intracellular and extracellular DNA, the former from living cells or multicellular organisms, the latter from dead and often ruptured and degraded cells (Taberlet et al., 2018). Extracellular DNA can be degraded and absorbed by inorganic or organic particles such as clay, sand, silt and humic substances (Taberlet et al., 2018). This is relevant because DNA fragmentation and absorption on to other substances can cause practical issues for eDNA methods.

There are two main approaches to biodiversity studies using eDNA, one is where the aim is to determine the presence or not of a single species generally based on PCR or quantitative PCR (Q-PCR). An example of this is the detection of dinoflagellates in seawater that can cause harmful algal blooms, which can lead to human health problems such as paralytic shellfish poisoning (e.g. detection of *Alexandrium* species; Vandersea et al., 2017), or, more recently, the early detection of outbreaks of corallivorous seastar (*Acanthaster* cf. *solaris*, Uthicke et al., 2022). Alternative approaches to single species detection using molecular approaches include fluorescence in-situ hybridization on fixed cells, sandwich-hybridization assays on lysed cells and other assays based on DNA amplification (e.g. quantitative or Q-PCR; Noyer et al., 2015). Microarray-based methods can be used to detect multiple species from PCR-amplified DNA collected from the environment (e.g. Phytochip for identification of HAB-forming algae; Noyer et al., 2015). The second approach using eDNA is also based on targeted

PCR or shotgun sequencing (see below) in order to identify the presence of multiple species in a sample. These methods are collectively termed DNA metabarcoding, although here we distinguish shotgun sequencing as a distinct method that will be covered in the next section.

DNA metabarcoding as usually practised today involves the PCR amplification of a specific microgenomic region (as in DNA barcoding) across the multiple fragments of DNA contained in an environmental sample (Brandt et al., 2021; Taberlet et al., 2018). When first used to identify the diversity of bacterioplankton in the Sargasso Sea in 1990 this PCR step was followed by cloning of the multiple PCR amplicons for subsequent sequencing (Giovannoni et al., 1990). This first metabarcoding study revealed the presence of the SAR11 cluster of α-Proteobacteria, which were subsequently discovered to be the most abundant group of bacterioplankton in the ocean (Morris et al., 2002). However, the cloning step required for isolation of single DNA fragments was both time consuming and expensive (Taberlet et al., 2018). It was the advent of next-generation sequencing in 2005 that provided the technological advance by avoiding the need for cloning that stimulated DNA metabarcoding. Three second generation sequencing technologies were developed: Roche 454 pyrosequencing, Personal Genome Machine or PGM (ion-torrent) and Illumina (Taberlet et al., 2018). Of these, Illumina sequencing has come to dominate in DNA metabarcoding. These sequencers can amplify 250–300 base-pairs (bp) of sequence and generate data at 800,000 times the rate of first-generation DNA sequencers (Taberlet et al., 2018). Third generation sequencers are now also available that can sequence single DNA molecules without prior amplification (Taberlet et al., 2018). These technologies include the PacBio Sequel system which uses Single Molecule Real Time sequencing technology (https://www.pacb.com/) and the Oxford Nanopore MinION technology (https://nanoporetech.com/). These technologies have a relatively high error rate in sequencing (~10%), which renders them inappropriate for DNA metabarcoding at present (Taberlet et al., 2018), although they can be used for shotgun sequencing and Deep-Learning tools are now being developed to support more utility (e.g. ONcorrect error correction tool which is reported to bring accuracy to 98.9–99.6%; Sahlin and Medvedev, 2021). It is likely that these technologies will continue to improve in the near future and the MinION system is portable and so can be used in the field. Indeed, Oxford Nanopore MinION technology has been used on board the RV *Ocean Explorer* in the recent Red Sea Decade Expedition. Full details of DNA metabarcoding approaches using Illumina sequencing are given in Taberlet et al. (2018). The approach involves designing PCR Primers for specific taxa (these can be as broad as Archaea, Bacteria or Eukaryota) which also have adaptor sequences included in them to enable the Illumina sequencing. A challenge with DNA metabarcoding is that it generates enormous quantities of data, which are a significant computing and storage challenge (Taberlet et al., 2018). Data require filtering for quality, reconstituting of sequences, sequence demultiplexing and dereplication prior to classification and identification (Taberlet et al., 2018).

eDNA approaches appear to be very effective in identifying the presence of invasive (e.g. invasive bivalve molluscs, Ardura et al., 2015; invasive fish Hatzenbuhler et al., 2017) or threatened species (Boussarie et al., 2018; Pikitch, 2018) as well as those which are very difficult to detect using other approaches (e.g. giant squid, Wada et al., 2020). It offers advantages over more conventional means for sampling threatened species including: better detection of species presence, assessment of distributional range and the diversity of associated biological communities, preventing injury to organisms or damage to their environment and also in terms of the lack of need for taxonomic expertise, as well as practicality and costs of locating what may be patchily distributed and rare individuals (Pikitch, 2018).

DNA metabarcoding is also being successfully applied to study the functional biodiversity of entire communities and revealing some expected and unexpected patterns of biodiversity (e.g. Obiol et al., 2020). Studies on coral reefs using Autonomous Reef Monitoring Structures indicate that the taxa most commonly observed in conventional morphology-based studies of reefs are the same as those detected using DNA metabarcoding based on COI (Arthropoda, Annelida and Mollusca most diverse operational taxonomic units [OTUs] with Chordata being very abundant; Carvalho et al., 2019; Pearman et al., 2020). Whilst a small number of OTUs were relatively abundant, both within and across reefs, the majority (78%) of OTUs represented rare species, a common observation in morphological studies of benthic communities (Carvalho et al., 2019). In both studies of Red Sea reef biodiversity, a large proportion of OTUs (45% in Carvalho et al., 2019) could not be assigned to any eukaryotic phylum (Carvalho et al., 2019; Pearman et al., 2020). Although some of these sequences may have resulted from artefacts in the DNA metabarcoding approach (Flynn et al., 2015), it also suggests that previous morphology-based studies have missed a significant number of species but also higher taxa, which probably occur in the very smallest body-size fractions (Carvalho et al., 2019).

Metabarcoding has also been applied to assessment of the biodiversity of sediment communities in seagrass beds. In this case both COI and 18S ribosomal RNA-encoding regions were amplified to reduce the effects of biased taxon sampling using a single region (Cowart et al., 2015). As with the coral reef studies a very large number of OTUs (12,948 or 93%) were unassigned to recognised metazoan groups. For COI the vast majority of these sequences corresponded to Archaea and Bacteria although 27.8% were eukaryotic sequences and 0.7% unassigned. For the 18S, 1,316 OTUs were identified of which 1,174 were metazoans with 94 coming from non-metazoan taxa and 48 unassigned, about half of which were identified as non-metazoan eukaryotes (Cowart et al., 2015). COI recovered Annelida, Arthropoda and Mollusca as the most frequent OTUs whilst the 18S recovered Annelida and Arthropoda but a lower proportion of Mollusca (Cowart et al., 2015). 18S was better at recovering Nematoda and Porifera, whilst COI and morphology were better at recovering Chordata and Echinodermata (Cowart et al., 2015). Molecular approaches recovered a much higher diversity of OTUs (5–10 times) than the morphological approaches (Cowart et al., 2015). However, the metabarcoding only detected 36% of the species which have been recorded from the study sites using morphology. This suggests that either the primers used for amplification of eDNA mismatched many of the common taxa or that sequences obtained for reference sequence databases from samples preserved in formalin may have been degraded (Cowart et al., 2015).

Deep-sea ecosystems including sediment associated with continental slopes, abyssal plains and canyons (Atienza et al., 2020; Brandt, 2020; Cordier et al., 2022; Guardiola et al., 2015, 2016), deep-sea hydrothermal vents (Cowart et al., 2020) and manganese nodule fields in the Clarion-Clipperton Fracture zone (Cordier et al., 2022; Laroche et al., 2020; Lejzerowicz et al., 2021) have also been investigated using DNA-metabarcoding. All of these studies have revealed similar findings to those in shallow water. These include a large number of unidentified OTUs or OTUs that have been clustered with known sequences but for non-deep sea or non-vent organisms (e.g. Cowart et al., 2020; Laroche et al., 2020; Lejzerowicz et al., 2021) or have been unassignable (e.g. Cordier et al., 2022). The former mainly results from a lack of coverage in current barcoding databases such as BOLD (e.g. Cowart et al., 2020; Laroche et al., 2020; Lejzerowicz et al., 2021). The latter has been suggested to arise from currently undescribed higher taxa located in deep-sea communities for which DNA sequence information has not been assigned

(Cordier et al., 2022). Use of different primer sets (i.e. COI, 18S) tended to favour amplification and subsequent recording of specific taxa (e.g. Atienza et al., 2020; Cowart et al., 2020; Laroche et al., 2020). All the studies demonstrated that DNA metabarcoding was able to demonstrate significant differences in assemblages related to habitat or other aspects of environmental variation as with the shallow-water studies on coral reefs and seagrass beds (Carvalho et al., 2019; Cordier et al., 2022; Cowart et al., 2015, 2020; Laroche et al., 2020; Lejzerowicz et al., 2021). Levels of biodiversity identified by DNA metabarcoding in deep-sea studies, especially those from the Clarion Clipperton Fracture Zone were comparable to diversity in seagrass beds and coral reef ecosystems. The overall identification of heterogeneity in the biodiversity of communities in the Clarion Clipperton Fracture Zone is particularly interesting in terms of management of deep-sea mining.

DNA metabarcoding has been identified as a potentially useful approach to biomonitoring surveys as a way of assessing ecosystem health (e.g. Aylagas et al., 2016, 2018, 2020; Cowart et al., 2020; Pawlowski et al., 2018; Valentini et al., 2016). Whilst biomonitoring of ecosystems using traditional approaches has always faced the challenge of species identification (hence the focus on indicator species) the use of DNA metabarcoding has been recognised as having a number of limitations (Pawlowski et al., 2018), some of which have already been identified above. These issues include overestimation of biodiversity, disagreement between molecular and morphological assessment of species that are present or absent and the estimation of abundance of taxa from molecular data. The issue of overestimation can arise from a variety of issues, especially in poorly studied groups of invertebrates, including high intraspecific variation, intragenomic variability and the occurrence of cryptic species (Pawlowski et al., 2018). One of the significant sources of error here is the application of a single threshold that arbitrarily identifies the level of genetic variation expected in intraspecific and interspecific comparisons. A solution may be the application of algorithms that generate thresholds for a network of closely related reads, or which take account of the distribution of sequences across reads (Pawlowski et al., 2018).

It should be noted that biodiversity can also be underestimated by DNA metabarcoding approaches, as a result of primer bias (Pawlowski et al., 2018), unequal biomass in samples between different size classes of organisms (Elbrecht et al., 2017), recent speciation (Pawlowski et al., 2018) and limitations of sampling in the barcoding process for rare species (Leray and Knowlton, 2017). To some extent the issue of unequal biomass can be compensated for by fractionating samples by size and normalising the amount of DNA from each size fraction which is used for metabarcoding (Elbrecht et al., 2017). This may be especially important in marine communities where there are a high proportion of rare species. Disagreement between molecular and morphological studies, as outlined above, is often attributable to the lack of data in molecular barcoding databases (Pawlowski et al., 2018). This might be addressed by the development of biotic indexes for DNA metabarcoding data, which do not refer to morphological taxonomy data at all (see Pawlowski et al., 2018). It can also be attributable to limited sampling of communities but can also be caused by the occurrence of 'ghost MOTUs', organisms represented by extracellular DNA that has remained in the environment after the organism has gone (Pawlowski et al., 2018). PCR primer bias can also contribute to this issue (Pawlowski et al., 2018). The sampling issue is one which may be addressed through use of increased sample replicates and stringent experimental procedures to reduce contamination of samples (Pawlowski et al., 2018). In terms of measuring relative abundance of taxa, some studies have shown that there is an abundance signal in the frequency of how often specific MOTUs are recovered in DNA metabarcoding (e.g. Lejzerowicz et al., 2015). However, because of the differential biomass of organisms in samples and also primer bias this remains a very difficult area in DNA metabarcoding (Pawlowski et al., 2018).

Despite the issues outlined above, DNA-metabarcoding does offer the possibility of biomonitoring with the inclusion of many taxa usually missed by morphological studies, especially small size fractions belonging to the meiofauna or species which inhabit cryptic habitats (e.g. Lejzerowicz et al., 2015; Pawlowski et al., 2018; Staehr et al., 2022). In a study of community diversity of subtidal boulder reefs in the Kattegat to the Baltic Sea diver surveys were combined with eDNA sampled from water collected near the seafloor (Staehr et al., 2022). The study demonstrated that whilst eDNA sequences missed some taxa (e.g. some macroalgae) and showed a higher level of dominance of some taxa than the diver surveys they also picked up markedly more fish species and also many more infaunal species than diver surveys (Staehr et al., 2022). The two methods applied together provided a more comprehensive picture of boulder reef community diversity and its changes across an environmental gradient (Staehr et al., 2022).

A unique capacity of eDNA is to provide a window to past biodiversity, not just extant one. This has been achieved by using sediment archives, which preserve eDNA over millenial times scales, coupled with suitable techniques to establish sediment chronologies, such as ^{210}Pb concentration profiles (Wesselmann et al., 2021, 2022). This approach allows the reconstruction of past changes in biodiversity (Wesselmann et al., 2022), can infer the timing of invasions (Wesselmann et al., 2021), and provide baselines of species abundance, which can be used to guide conservation targets.

Despite the advances in species detection and discovery (of M-OTUs) using metabarcoding approaches a major challenge in biodiversity assessment in the ocean is it sheer size and the remoteness of many ecosystems. Use of ship-based sampling whether direct (e.g. use of Niskin bottles for water sampling) or indirect (e.g. through deployment of remotely operated vehicles or submersibles with Niskins) will be a limiting factor in the sampling of eDNA. Another consideration is the CO_2 emissions impacts from the use of ocean-going vessels. The use of automated platforms is an obvious solution to larger-scale sampling of the ocean including remote ecosystems. The Environmental Sample Processor (ESP) is a device which can collect multiple environmental samples from seawater either for subsequent metabarcoding studies or for onboard detection of specific taxa or genes using sandwich hybridisation assays or Q-PCR (Scholin et al., 2017). The ESP has been termed an ecogenomic sensor and has now been mounted on a Tethys Long Range Autonomous Underwater Vehicle (LR-AUV) enabling it to take multiple samples of eDNA for the detection of a wide range of organisms including microbial taxa, zooplankton and animals (e.g. Scholin et al., 2017; Yamahara et al., 2019; Zhang et al., 2019). At present the undertaking of DNA sequencing is too complex for application within the ESP or similar device mounted on an AUV (Scholin et al., 2017) but rapid technological advances in miniaturised sequencing technology may make this feasible in the near future. This could be transformative to our understanding of the spatial and temporal patterns of biodiversity present in marine ecosystems.

To conclude, there is little doubt that DNA metabarcoding approaches can provide a new perspective on marine biodiversity either being deployed singly or as a complementary approach with other forms of survey. This method can give insights into how communities vary over space and time, with clear applications in ecology and applied biomonitoring (see references above). It may allow for the first time the ability to study patterns in the distribution of many groups of small organisms or those which inhabit cryptic habitats which are difficult to sample and identify using conventional methods (e.g. Creer et al., 2016; Staehr et al., 2022). However, DNA metabarcoding approaches still have limitations. Some of these can be overcome by careful application of methodologies, particularly in terms of careful pilot studies and consideration of both sampling strategies and data analyses approaches. There is clearly an urgency to both populate and curate DNA barcoding/metabarcoding databases to improve the utility of these approaches. Some limitations to DNA metabarcoding methodologies may be resolved through further developments of the approach, others will remain and their implications for biodiversity assessment and monitoring understood so that data may be correctly interpreted. The possibility of combining eDNA approaches with autonomous platforms is a promising area of future research that may open up remote ocean areas for biodiversity assessment and monitoring. DNA sequencing opens, eventually, the opportunity to derive taxonomy-free biodiversity assessments, thereby freeing biodiversity science from the constraints associated with taxonomic assessments, but which require stringent methodologies to avoid overestimates and artifacts (e.g. Mächler et al., 2021).

3.3.4.4 Genome sequencing (metagenomics; shotgun sequencing)

Metagenomics is the generation of long-sequence or genomic reads of DNA for pooled individuals or from community samples (Krehenwinkel et al., 2019a). Reads can be analysed and compared to databases or they can be assembled into partial genome sequences (Krehenwinkel et al., 2019a). Alternatively, only a subset of the reads in the genomic library are used, such as genes located in mitochondrial and chloroplast genomes for species identification (genome skimming; Krehenwinkel et al., 2019a). Compared to conventional DNA metabarcoding, this approach can produce much longer and phylogenetically more informative sequences than the short reads produced by Illumina sequencers (Krehenwinkel et al., 2019a). Increasing sequencing depth on metagenomic studies is also now allowing the assembly of high-quality full genome sequences for marine microbes, termed metagenomic-assembled genomes (MAGs), providing advanced sequencing information on organisms that have never been cultured or formally described (Parks et al., 2017).

The advent of third generation sequencers, especially the Oxford Nanopore Technologies MinION sequencer, has made biodiversity assessment using metagenomic approaches more economical. These sequencers are also portable and can be used in the field reducing the need for the long process of sample collection, preservation, transport from the field, potentially across jurisdictions (thus requiring permitting; e.g. Pomerantz et al., 2018) and then sequencing in molecular laboratories (Bleidorn, 2016; Krehenwinkel et al., 2019a,b; Maestri et al., 2019; Menegon et al., 2017).

Metagenomic and shotgun sequencing technologies have mainly been applied to communities of marine microorganisms. The most notable examples of this have been from the *Tara* Oceans (Karsenti et al., 2011) and *Malaspina* Expeditions (Duarte, 2015), which have sampled microbial communities from the ocean surface to the deep sea respectively. Environmental samples were size fractionated on board the sailing vessels *Tara* and *Hesperides* preserved specifically for genomics studies and returned to land for sequencing of genomes (including from single cells) and genes expressed within planktonic communities (Acinas et al., 2021; Alberti et al., 2017; Karsenti et al., 2011; Pesant et al., 2015; Seeleuthner et al., 2018). The impressive scientific knowledge gathered by this programme included the fact that more than 80% of the genes discovered from the sampled organisms were previously unknown to science (Sunagawa et al., 2015). The genomic information also revealed previously undescribed groups of microbial eukaryotes, as well as large numbers of undescribed species from known groups of organisms (De Vargas et al., 2015). Together these programmes have tripled the number of known viral populations from the ocean and doubled the number of bacterial and archaeal virus genera (Roux et al., 2016). This number was later expanded 12-fold (195,728 viral populations) through more in-depth sequencing of mesopelagic samples (zone between 200m and 1,000 m depth) and inclusion of samples from the *Tara* Oceans Polar Circle expedition (Gregory et al., 2019). The analyses of bathypelagic microbial metagenomes from the *Malaspina* Expedition formed the basis for the recently published *Malaspina* Deep-Sea Gene Collection. The collection consists of mostly novel genes (71%), and its analysis has uncovered 11 potential new phyla, greatly expanding our understanding of the functional diversity and metabolic versatility of deep-ocean microbial taxa (Acinas et al., 2021). *Tara* Oceans also identified 116 million genes expressed by planktonic organisms across the global ocean (Carradec et al., 2018; see also Salazar et al., 2019). Replicated metagenomic sequencing of marine microbial communities showed that novelty continues to appear, even at high protein family level, with increasing sequencing, suggesting that the current gene inventory for the ocean is far from complete (Duarte et al., 2020a,b).

The programme implemented an open data archiving strategy from its inception and nucleic acid sequence data were deposited in the International Nucleotide Sequence Database Collaboration (INSDC, http://www.insdc.org/about) via the European gateway to the INSDC, the European Nucleotide Archive (ENA, http://www.ebi.ac.uk/ena) at the European Molecular Biology Laboratory's European Bioinformatics

Institute (EMBL-EBI; Alberti et al., 2017). All environmental and biogeochemical measurements for each *Tara* Oceans sample were deposited at the PANGAEA database (https://www.pangaea.de/).

The application of metagenomic and shotgun-sequencing based approaches to larger organisms has to date been very limited. For instance, metagenomic data from the Tara and Malaspina expeditions has been assembled to resolve the presence of macroalgal genes across the open and deep ocean (Ortega et al., 2019). However, the increasing number of publications in this area suggest that these technologies are becoming more widely adopted in biodiversity studies. Of the studies undertaken to date, most have used 3rd generation sequencing technologies to sequence multiple PCR amplicons of barcoding regions. This method gets around the high error rates of MinION sequencing by analysing data to produce consensus sequences from multiple amplicons, each of which will have random errors as a result of the MinION technology. Applications to date have included identification of coral hosts, symbionts and microbiomes (Carradec et al., 2020), identification of harmful algal bloom-forming diatoms (Hatfield et al., 2020), identification of jellyfish (Ames et al., 2021), assessment of biodiversity on coral reef ARMs devices (Chang et al., 2020) and identification of invasive species (Egeter et al., 2020). True shotgun sequencing (metagenomic) studies have employed samples of DNA or RNA, the advantage of the latter being that it represents only living organisms rather than extracellular or ghost DNA (Broman et al., 2020; Cowart et al., 2018; Semmouri et al., 2019). These studies have used second (Broman et al., 2020; Cowart et al., 2018; Singer et al., 2020) or third (Semmouri et al., 2019) generation sequencing of generated libraries. All of these studies were able to identify species from consensus sequences. They also had the advantage of being able to also identify functional genes, giving additional insights into how the environment influenced community function (Semmouri et al., 2019; Singer et al., 2020). A study that compared results of DNA metabarcoding vs the metagenomic approach for samples taken in nearshore and offshore locations in Newfoundland concluded that the former identified more taxa than the latter (Singer et al., 2020). However, the metagenomic approach seemed to identify a different set of organisms and, as with the DNA metabarcoding assessment, could identify community differences between samples (Singer et al., 2020). The two methods were therefore viewed as complementary (Singer et al., 2020). Most of these studies avoided the use of PCR amplification of samples or specific gene regions because of concerns about the introduction of bias.

Metagenomic approaches therefore provide a very powerful technology for not only assessing biodiversity but also gaining insights into ecosystem function of the species present. The adoption of Oxford Nanopore Technologies MinION sequencers means that it is possible to take molecular biodiversity assessment into the field. This lends itself to teaching both DNA metabarcoding and metagenomic approaches to students in low- and middle-income countries or in the field (Krehenwinkel et al., 2019a; Watsa et al., 2020). It is likely that, over the next few years, more studies will elucidate the best uses of metagenomic approaches for biodiversity research as well as its limitations.

Most metagenomic data are publicly available, as practitioners of molecular biodiversity research have adopted practices that require publication of the sequence data in open repositories. Hence, increasingly the bottle neck for participating in sequencing-based biodiversity research is no longer the access to sequencing instrumentation or funding for sequencing, but the access to supercomputers and advanced bioinformatic capabilities required to access and analyse the big data sequencing, particularly metagenomics, represent. An important development is the advent of open metagenomic analysis platforms, such as KMAP (Alam et al., 2021), which allow researchers to query large taxonomically-resolved metagenomic data sets from a personal computer connected to the internet. However, these platforms are, for the time being, only able to analyse microbial DNA sequences.

3.3.5 Acoustic approaches

3.3.5.1 Multibeam and multibeam backscatter data

Prior to the 20th century the depth of the ocean was measured using the lead and line system in which a plumb attached to a cable is allowed to sink to the seafloor (Menandro and Bastos, 2020). The history of echosounders began with the *Titanic* disaster of 1912 after which a patent was submitted by LF Richardson for an air and underwater echo-ranging apparatus for detection of icebergs (Hersey and Backus, 1962). At the same time RA Fessenden in the USA was working on a moving coil transducer for underwater echo-ranging and signalling which, in 1914 was used to detect an iceberg at a range of 2 miles (Hersey and Backus, 1962). These systems were developed during World War I for detecting submarines although they came too late to have much of an effect on the war. Adaption of this sonar apparatus for measuring water depth was very straightforward involving orientation of a directional transducer vertically downwards (Vigoureux and Hersey, 1962). This transducer emits regular pulses of sound so that between pulses it can serve as the receiver of the bottom echo (Vigoureux and Hersey, 1962). A graphic recorder for collecting echosounder data was introduced by Marti (1922) and these instruments were then regularly used for measuring water depth. The German Meteor Expedition (1925–1927) surveyed the South Atlantic Ocean using echo sounding equipment and other oceanographic tools (Menandro and Bastos, 2020). This expedition was notable for identifying the continuous nature of the Mid-Atlantic Ridge (Menandro and Bastos, 2020). In the 1950s and 1960s improvements were made on sonar technology including electronic stabilization, interferometric modulation and positioning (Menandro and Bastos, 2020). Multibeam bathymetry systems were developed in the 1960s by the military for seafloor mapping related to submarine warfare. These systems emit a fan-shaped wave of acoustic pulses (>100 sound beams and in some cases >200; Brown and Blondel, 2009; Harris and Baker, 2012a) which are reflected from across a swath of the seafloor and picked up by the transceiver. They revolutionised seafloor mapping by enabling a broad track of the seafloor to be mapped under a vessel. The first civilian systems were installed on the French research vessel *Jean Charcot* and employed on a first expedition in 1977. Multibeam sonar when combined with Geographic Positioning System (GPS data) produces high resolution geopositioned contour maps of the seafloor (Harris and Baker, 2012a). However, it was discovered that the properties of the backscatter data from such echo sounders (e.g. the strength of the return echo) were dependent on seafloor hardness, texture and roughness (De Moustier, 1986). Different frequencies of echosounders are used to map different water depths with higher frequencies used for shallower water mapping (see below; Harris and Baker, 2012a). Sidescan-sonar only collects acoustic backscatter data but because it is usually deployed as a tow-fish behind a ship it is not easy to exactly georeference the map produced. Therefore, despite it being less expensive than multibeam systems and easier to deploy from smaller vessels it has significant disadvantages for habitat mapping (Harris and Baker, 2012a). Approaches to analysis of backscatter data are discussed in Brown and Blondel (2009). This method is now improving with the advent of multifrequency acoustic backscatter data (e.g. Brown et al., 2019; Costa, 2019; Feldens et al., 2018).

Both seafloor topography and backscatter data give important information on benthic habitat, the first for identifying important topographic features (e.g. seamounts, ridges, canyons) and the second on the nature of the seafloor (e.g. basalt, sand, coral). Both of these types of data provide information on physical habitats, the ecological or environmental areas that are inhabited by particular species (Harris and Baker, 2012a). Importantly, the same types of habitats are consistently associated with biological communities or assemblages of species which occur together (Harris and Baker, 2012a). An example would be the occurrence of rich coral communities associated with seamounts (e.g. Rogers, 1994, 2018) or kelp forest communities with shallow, wave-influenced rocky seabed in temperate and sub-polar latitudes (Harris and Baker, 2012a). Seascapes comprise suites of habitats that consistently occur together (the marine equivalent of landscapes; Harris and Baker, 2012a). The classification of habitats is structured hierarchically reflecting increasing degrees of similarity (see Harris, 2012a for review). Currently, it is still the case that our ability to map the physical structure of the seabed greatly exceeds our ability to map what species live in these habitats (Harris and Baker, 2012a). Therefore, mapping of physical habitats is known as 'potential habitat mapping' as there are huge geographic gaps in our observations of life on the seafloor (Harris and Baker, 2012a). What can be achieved instead is to carry out biological observations (e.g. sampling of biological communities using video survey or other methods) on a small subset of the seafloor for which acoustic data has generated topographic maps and backscatter data and then to use these data to model community distribution at much larger geographic scales. Such habitat maps, whether developed from observations and/or modelling have a range of important uses including (after Harris and Baker, 2012a):

- Support of marine spatial planning, ocean management and decision making
- Support and underpinning of the design of marine protected areas
- As part of scientific research on benthic ecosystems and seafloor geology
- For assessment of biotic and abiotic seafloor resources

See Baker and Harris (2012) for a more in-depth discussion of the uses of habitat mapping data.

The mapping of biophysical variables of the seafloor which have a quantifiable correspondence to the occurrence of benthic species and communities is referred to as surrogacy (Harris, 2012b). Surrogacy is an empirical method of determining which measured seafloor characteristics can be used to best describe the occurrence of a species assemblage in a particular space (and potentially time; Harris, 2012b). These characteristics can be used as predictors for the occurrence of species assemblages in areas which have been physical mapped but not subject to biological observations through the use of modelling. A range of geomorphological features are described in Harris (2012c).

Whilst Harris (2012c) details large-scale habitat features there are other ways of classifying physical habitat from seabed mapping data. Benthic Terrain Modelling (BTM) is probably one of the most widely employed approaches based on tools available with Geographic Information System (GIS) software (e.g. Taylor et al., 2013; Wilson et al., 2007b). Examples, of these approaches to habitat mapping are given in Harris and Baker (2012b). Generally, two main approaches are used for habitat mapping, supervised and unsupervised classification (Calvert et al., 2015). In supervised classification ground truth data is interpreted and used to constrain the interpretation of acoustic data (Calvert et al., 2015). In unsupervised classification data are clustered or classified by pattern recognition and are assigned habitat type based on the spatial co-occurrence of classified map pixels and ground-truth data (Calvert et al., 2015). Machine Learning approaches are now being applied to physical habitat classifications (Pillay et al., 2020; Sture et al., 2018). Processing methods for multibeam data from the water column have also now been used to identify habitats such as kelp forests (e.g. Schimel et al., 2020) as well as gas plumes from hydrocarbon seeps (e.g. Von Deimling et al., 2007; Wilson et al., 2015). A number of other technologies such as satellite remote sensing and Laser-Light Detection and Ranging (LIDAR) can also be used for mapping shallow-water or coastal habitats (Harris and Baker, 2012a). A new development of side scan sonar known as Synthetic Aperture Sonar (SAS) has also been used in conjunction with Convolutional Neural Networks (CNN) to identify coral habitat (Sture et al., 2018).

Habitat classification and mapping is an important area of biodiversity research with many applied and scientific applications. The combination of further refinements in acoustic mapping technologies (e.g. multi-frequency multibeam bathymetry; synthetic aperture sonar) and navigation together with application of new methods such as Deep Machine Learning to habitat classification are increasing the resolution and accuracy of acoustic mapping of habitat.

3.3.5.2 Active biological acoustics

The first use of active acoustics to study pelagic ecology occurred in the 1920s and 1930s with Kimura (1929) being the first scientist to record echoes from a fish (Benoit-Bird and Lawson, 2016). In about 1930 the pioneering British fisherman Captain Ronald Balls installed an echosounder in his herring drifter *Violet and Rose* and was able to set his nets on mid-water echoes correctly attributed to herring shoals (Hersey and Backus, 1962;

Balls' observations were recorded in two papers, Balls, 1948, 1951). Early work on the use of echosounders was also undertaken in Norway with the plotting of echoes in a time-depth plot (Sund, 1935). During World War 2 sound propagation experiments undertaken by workers at the University of California's Division of War Research identified a consistent layer of mid-water sound scatterers which migrated over a diurnal period to the surface and back to depth again (Hersey and Backus, 1962). This became known as the 'deep-scattering layer' (Johnson, 1948). Although the origin was concluded to be biological, the identity of the organisms reflecting sound energy remained elusive for many years because of difficulties in sampling deep-water pelagic ecosystems (Hersey and Backus, 1962).

The principle behind biological acoustics is that sound pulses are produced and their echoes from living organisms in the water column are received by a transducer mounted in the hull of a vessel or autonomous platform or towed behind a vessel in a profiler or mounted on a mooring (Jennings et al., 2001; Benoit-Bird and Lawson, 2016; Fig. 8). When a sound pulse meets an object with a density different from seawater sound waves are scattered in all directions (Jennings et al., 2001). Since they spread spherically their intensity decreases rapidly with distance so the signal must be magnified by a gain function in the echo-sounding equipment (Jennings et al., 2001). Early sounders recorded single echoes, but the development of the echo-integrator allowed the strength of many echo returns to be summed over depth and distance (Jennings et al., 2001). The advent of multifrequency echosounders, multiple echosounders emitting sound pulses at different frequencies or a single echosounder with multiple processing units (Korneliussen, 2018) has been critical in the ability of acoustics to discriminate among species and taxa (Benoit-Bird and Waluk, 2020). This has been further developed with the use of broadband echosounders that transmit and receive a frequency-modulated signal from a single transducer to allow acoustic backscattering to be measured continuously over a range of frequencies (Benoit-Bird and Waluk, 2020). Combining multiple broadband measures to develop wideband responses may further increase the information available for the characterization and classification of biological targets (Benoit-Bird and Waluk, 2020). Different types of organisms have different sound reflection properties, for example, those with gas-filled cavities such as siphonophores or fish with swim bladders have a resonant sound scattering frequency that depends on the depth and size of the cavity (Benoit-Bird and Lawson, 2016; Benoit-Bird and Waluk, 2020; Korneliussen, 2018; animal groups 5 and 7 in Fig. 9). Elastic shelled zooplankton, such as krill show

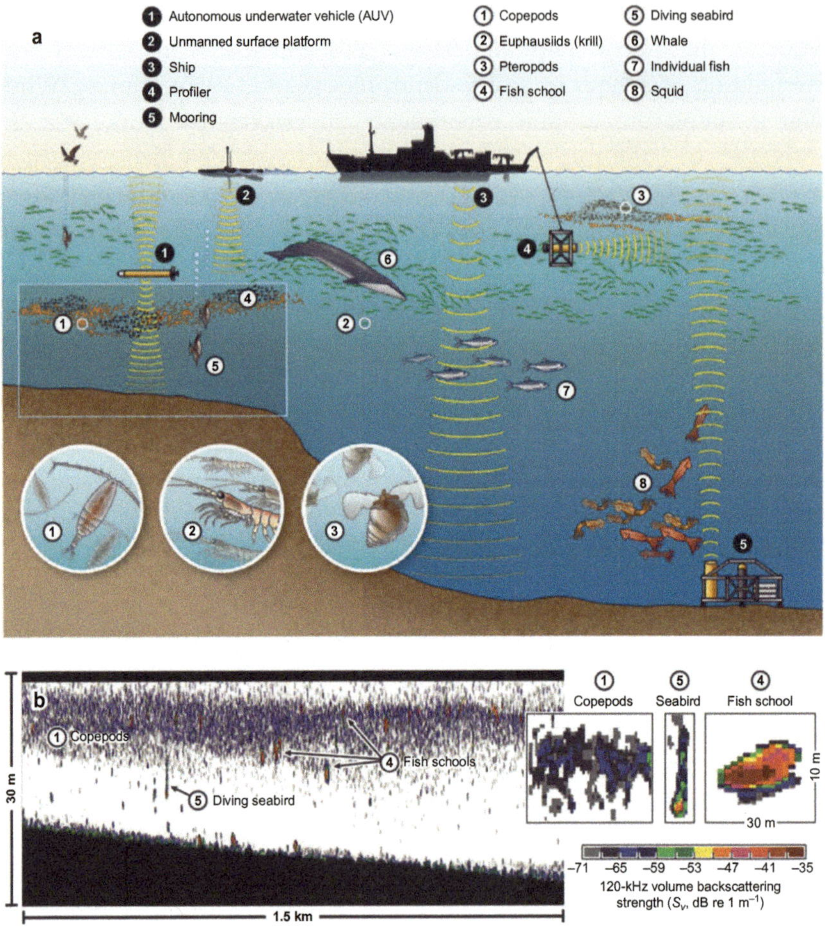

Fig. 8 Graphic showing the range of platforms that can carry active acoustic equipment and the range of organisms that can be detected using it (Benoit-Bird and Lawson, 2016).

a smooth transition in reflection between low and high sound frequencies (Korneliussen, 2018; animal group 2 in Fig. 9). Fluid-filled organisms with a sound-speed and density similar to seawater have a backscatter that fluctuates between low and high frequencies (Korneliussen, 2018; animal groups 1, 2 and 3 in Fig. 9). The aim of acoustic target classification is to categorise sound scatter into groups which then hopefully correspond to different species or groups of organisms for which abundance and biomass can then be estimated (Korneliussen, 2018). The process of target classification involves (i) discrimination (removal of unwanted data from that to be processed)

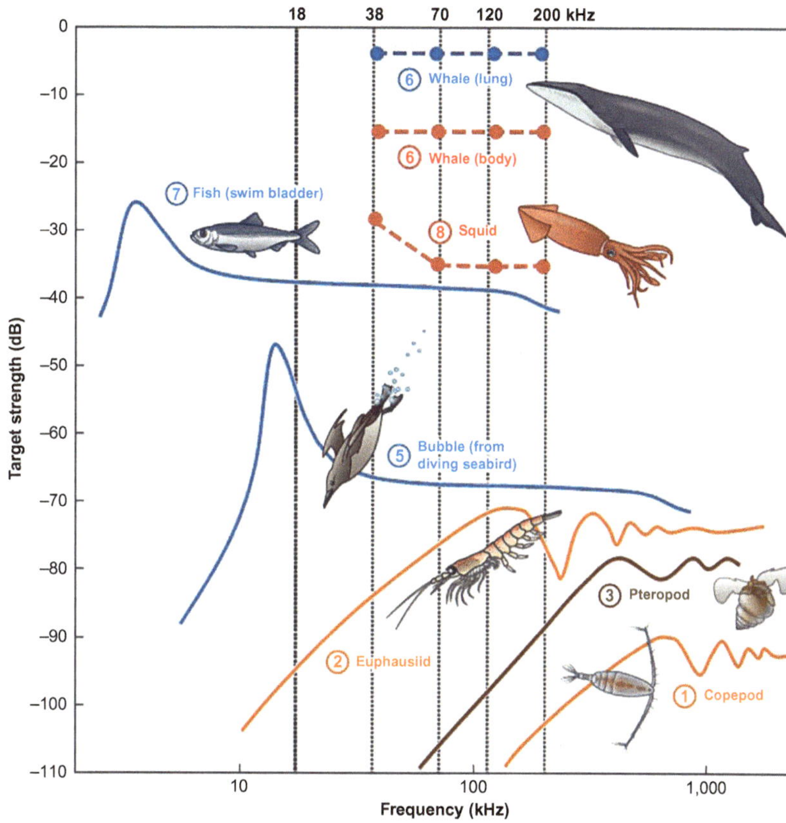

Fig. 9 Frequency responses characteristic of the different types of animals shown in Fig. 8, utilizing the same circled numbers. Dashed lines denote curves derived from empirical observations; solid lines indicate predictions based on physics-based scattering models validated through empirical measurements. The vertical dotted lines show typical sampling frequencies (18, 38, 70, 120, and 200 kHz). The line colors indicate animal types (or parts of animals) that are similar kinds of acoustic scatterers (Benoit-Bird and Lawson, 2016).

(ii) categorization (grouping of scatterers with similar properties) (iii) identification (iv) validation (ground truthing of the data, for example with trawl or video survey data; Korneliussen, 2018). Categorization and identification are achieved through analyses of frequency responses in comparison to models or previous observations and the echo distributional characteristics, for example the shape or depth of aggregations and/or behaviour of the scatterers (Benoit-Bird and Lawson, 2016). If the target strength of the species under investigation is known then abundance and biomass can be estimated from acoustic data (Benoit-Bird and Lawson, 2016).

Approaches to classification of echograms is based on model-based approaches or on empirical approaches (use of prior-collected data on sound scattering properties of monospecific aggregations of a species; Korneliussen, 2018). If species aggregations are known to have specific sizes or shapes these can be used to identify species from the dimensions and shapes of echograms (echo envelope; Korneliussen, 2018). Multivariate statistical approaches have been used to classify species on the basis of echo envelopes, other information on aggregation shape and size and also physical environmental data (Korneliussen, 2018). Increasingly, Artificial Neural Networks are being applied to the identification of species (e.g. Brautaset et al., 2020; Korneliussen, 2018). Classifications, like those for habitat mapping, can be supervised or unsupervised (Korneliussen, 2018).

The main scientific uses of active biological acoustics have been the study of where animals occur in the ocean and what drives their distribution (Benoit-Bird and Lawson, 2016). Acoustic studies are also useful at understanding the temporal dynamics of animal populations over a range of scales (Benoit-Bird and Lawson, 2016). They can also be used to understand the effects of climate change and other human impacts on pelagic ecology as well as on the physiology of animals (Benoit-Bird and Lawson, 2016). Other aspects of ecology that can be studied include life-history strategies, how animals respond to environmental stimuli and predator-prey relationships (Benoit-Bird and Lawson, 2016). Active acoustics are an essential element of biodiversity studies on pelagic organisms as well as the interaction of such organisms with benthic ecosystems. Further developments in both echosounders (e.g. broadband echosounders; Benoit-Bird and Waluk, 2020) and data analyses (e.g. Brautaset et al., 2020) are improving the discriminatory abilities of acoustic approaches to study biodiversity. The combination of traditional sampling methods, new approaches such as eDNA (e.g. Easson et al., 2020) and acoustics can be further integrated to improve pelagic biodiversity studies even further.

3.3.5.3 Passive acoustics

Because sound is rapidly and efficiently transmitted underwater and over great distances it is of no surprise that it is often used by animals for communication or produced by them as a result of other activities (Mooney et al., 2020; Schevill et al., 1962). The writings of Aristotle, Athanaeus and Pliny describe sound production from whales, fish and crustaceans from ancient times (Schevill et al., 1962). Since ancient times shoals of fish have been located underwater by listening with the unaided ear a practice which continued into modern times by Indigenous peoples and Chinese fishers (Sadovy and Cheung, 2003; Schevill et al., 1962). During the 19th Century scientists began to investigate species of fish which could produce sound, investigating the anatomical sound-making structures and speculating as to function (Schevill et al., 1962). Again, the technological advances during World War 2 along with the military importance of understanding noise underwater led to significant advances in the science of biological sound underwater (Schevill et al., 1962). Hydrophones have been used in the study of underwater noise production from animals since at least 1940 (Schevill et al., 1962).

Conventional methods of biodiversity assessment rely on capturing organisms or on visual or image-based census. Such surveys only provide a picture of biodiversity at a few points in time, they miss rare or cryptic species (e.g. animals hidden in burrows, or cavities in rock; Mooney et al., 2020; Picciulin et al., 2019) and are not well suited to assessing populations of animals which spend large amounts of time submerged and range over very large areas (e.g. cetaceans; e.g. Marques et al., 2009). Passive acoustic monitoring (PAM) can be undertaken over long periods of time, in a range of weather conditions and in the day or night. This is assisted by a new generation of advanced passive acoustic recorders which allow broadband, long-term recordings and which are relatively cost effective (Mooney et al., 2020). The marine environment is rich in biological sound (biophony) being produced by a range of organisms during spawning, courtship, feeding, social cohesion and competition (Mooney et al., 2020). The acoustic signals of species generally have distinct characteristics such as frequency (pitch), amplitude, duration and repetition rate and it is assumed that they have limited overlap (Mooney et al., 2020). This raises the prospect of being able to monitor either the presence and behaviour of particular species or to monitor the overall biodiversity in the environment.

There are many examples of the use of passive acoustic monitoring to detect the abundance, distribution and behaviour of marine mammals (e.g. sperm whales, Morrissey et al., 2006; Blainville's beaked whale, Marques et al., 2009; killer whales, Hanson et al., 2013; North Atlantic right whale, Davis et al., 2017; Salisbury et al., 2015). More than 700 species of fish are also known to produce sound and so passive acoustic monitoring has been used to study migration, spawning patterns, behaviour, including courtship, intraspecific competition and aggression (Luczkovich et al., 2008). It has also been used to detect the presence of the rare and cryptic species of eel, *Ophidion rochei* in marine protected areas where visual census of fish populations over many years had failed to detect its presence (Picciulin et al., 2019). Monitoring the biodiversity of an entire community of organisms is a very different prospect to acoustic monitoring of one or a few species with known acoustic characteristics. Work in terrestrial ecosystems have suggested that the biological soundscape can be an indicator of levels of biodiversity (Mooney et al., 2020). Some marine studies have indicated that some types of biological noise can be related to biodiversity. For example, in Caribbean coastal habitats it has been found that acoustic measurements in the low frequency band (<1000 Hz) were positively correlated with the diversity and abundance of cryptic fish species (Staaterman et al., 2017). However, in many other cases such correlations have not been detected. A variety of issues have been identified that can interfere with the use of passive acoustic monitoring data as a proxy for biodiversity. These include: lack of sufficient spatial and temporal replication of data recording; masking of biological sound by anthropogenic or geophysical noise; masking of sound from elements of the biological community by loud and/or continuous noise (e.g. choruses) from other soniferous species and failure to capture transient noise (Mooney et al., 2020). Whilst various approaches to analyses of data from PAM devices for biodiversity monitoring are being developed, including Deep-Learning approaches, no standard approach for quantifying biodiversity acoustically has been adopted by the research community (Mooney et al., 2020). Further research and technology development are therefore required. It should be noted that there is no global database for ocean noise (Mooney et al., 2020) although there are for specific types of sound produced by ships (e.g. Santos-Domínguez et al., 2016) or by marine mammals (e.g. https://cis.whoi.edu/science/B/whalesounds/index.cfm) or noise libraries for specific geographic regions (e.g. Centre for Environment, Fisheries and Aquaculture Science, U.K.'s Sound Library). However, there has been a call for the development of an international platform for sharing acoustic data (Parsons et al., 2022). It is also important to note that the availability of low-cost PAM recorders lends the technology very well to citizen science projects and these would be greatly facilitated by a global platform for acoustic data sharing (Parsons et al., 2022; Radeta et al., 2018).

3.3.6 Biodiversity databases and related organisations

Databases or data platforms play several important roles in marine taxonomy and biodiversity research beyond simply storing data. They connect data producers (scientists) with users (e.g. other scientists, environmental managers, governments, the public) providing information that can be used for monitoring, assessment, projection and management of the marine environment as well as for public engagement. Databases relevant to marine species discovery and biodiversity include those focusing on cataloguing species names or taxonomic information (e.g. WoRMS), species occurrences (e.g. OBIS), genetic data (e.g. International Nucleotide Sequence Database Collaboration; INSDC), and data on the environment from where species are sampled (e.g. Pangaea). Some databases include information on a wide range of aspects of biodiversity including taxonomy, occurrence data and information on biology (e.g. Fishbase). These databases vary in size and also integration, with the genetic/genomic databases being large, well-supported and highly integrated whereas taxon-specific databases can be small and supported by a few experts (e.g. Brachiopod Database).

Overall, the picture is one of substantial data resources for marine biodiversity research and management but these databases are fragmented at a variety of levels and are generally tailored to specific user communities (usually scientists). For these databases to realise their full potential in terms of accelerating the discovery and description of marine species and biodiversity science greater connectivity and interoperability is required. Further analyses of global marine biodiversity databases and hosting organisations will not be undertaken in this paper because they are currently undergoing significant changes to improve their scope, interoperability and use for stakeholders. Many of these changes have not yet been captured by the literature. However, examination of the resources available suggest that rather than development of new marine biodiversity databases the emphasis of future development should be to make data more accessible for a wide variety of stakeholders so that it is translated into actionable knowledge. This may be a key foundation in integrated taxonomic approaches and cybertaxonomy (see Conclusions below).

3.3.7 Citizen science

Citizen science has existed since the late 1800's (Dickinson and Bonney, 2012) and arguably has origins that are a lot older (e.g. Johannes, 1981). It was initially coined to distinguish between professional and amateur scientists and includes a vast array of activities (see below). However, citizenship is not a qualification for participation in these programmes, and thus 'citizen' is now considered, by many, to be an exclusionary term. Also, citizens have different rights and responsibilities depending on the state, thus, there are different expectations conferred with the word 'citizen'.

Community science was initially suggested an alternative label to citizen science, and many more terms have since been suggested, with the most recent being 'tracking science' which has explicit inclusion of Indigenous peoples (Liebenberg et al., 2021). In this paper we have retained the term citizen science as it is the expression most widely used to characterise amateur science and there are some that do advocate for the retention of this term and instead to promote more inclusive practices (Cooper et al., 2021). However, as practices such as co-design and peer-to-peer learning become more widely used, further definitions are likely to evolve (Dosemagen and Parker, 2019) to match changing approaches.

3.3.7.1 Science contribution

More effective efforts are urgently required to improve the capacity of marine conservation action worldwide, as highlighted by the United Nations Decade of Ocean Science for Sustainable Development 2021–2030. Citizen science offers great potential for gathering scientific data over large spatial and temporal scales and for remote geographic regions and/or those difficult for professional scientists to reach and survey or monitor (e.g. Edgar et al., 2017b; Garcia-Soto et al., 2021; Gouraguine et al., 2019; Lauro et al., 2014; Norman et al., 2017). Projects have included a very wide range of topics including monitoring of pollution (e.g. plastic debris e.g. Van der Velde et al., 2017), monitoring of rare or threatened species (e.g. Edgar et al., 2017b; Norman et al., 2017), monitoring of marine protected areas (e.g. Freiwald et al., 2018) and monitoring of ecosystem biodiversity and health (e.g. Garcia-Soto et al., 2021; Gouraguine et al., 2019). A recent survey of citizen science projects in the EU identified species and biodiversity monitoring projects as by far the most common (Garcia-Soto et al., 2021). Citizen scientists have proved to be effective at data collection, sometimes as effective or even more effective than professional scientists (e.g. Delaney et al., 2008; Freiwald et al., 2018; Van der Velde et al., 2017). An obvious limitation of marine citizen science projects when compared to terrestrial projects is that there is limited accessibility to most of the ocean. As a result, the majority of projects are focused on beaches and other coastal ecosystems (e.g. Garcia-Soto et al., 2021). Some communities do, however, have access to submarine or more offshore habitats (e.g. SCUBA divers, yacht sailors and fisher folk) enabling projects in these realms (e.g. Edgar et al., 2017b; Freiwald et al., 2018; Lauro et al., 2014; Norman et al., 2017). Projects in offshore, polar and deep-sea ecosystems tend to be focused on data extraction from video or photographs for training of

Machine Learning algorithms (e.g. Anton et al., 2021; Jones et al., 2018b; Langenkämper et al., 2019). Developments in technology, especially smart phone-based applications for photo-based survey, self-build cheap sensors, aerial drones combined with AI, cloud-based and edge computing, social media, platforms, remote sensing and other areas mean that citizen science is becoming more capable of undertaking sophisticated scientific data collection tasks as well as more participatory projects in terms of coproduction of science (Garcia-Soto et al., 2021).

Such is the scale of the biodiversity challenge (e.g. only 10% of marine species have been described) that the harnessing of citizen science capacity has been identified as being critical to scaling up global efforts to describe species (e.g. Godfray, 2007). Conservation initiatives that collaborate with citizen scientists and involve local stakeholders can be more effective because they can engage local and traditional knowledge more readily, inspire conservation-minded behaviours, encourage relationship-building and increase the legitimacy of the science and data used (e.g. Dean et al., 2018; Freiwald et al., 2018). Citizen science is demonstrated to contribute to science knowledge gaps by using 'people power' to collect data in remote regions, in real time and in high quantities (e.g. Edgar et al., 2017b; Gouraguine et al., 2019; Norman et al., 2017). Also different types of information can often be more readily accessed and documented, supporting different way of observing and knowing a system. Data obtained through these projects can make valuable contributions to populating broad-scale databases and informing natural resource management and policy, including the potential to monitor change in relation to the UN Sustainable Development Goals (e.g. Garcia-Soto et al., 2021; Kelly et al., 2020). Participation in citizen science projects engages people with their local environments, enhances social capital and encourages stewardship and marine citizenship by providing 'hands-on' experiences that generate greater awareness of marine environments (e.g. Dean et al., 2018; Kelly et al., 2020).

3.3.7.2 Citizen science principles and practice

Several citizen science associations have been established in recent years in Europe (European Citizen Science Association, ECSA), Australia (Australian Citizen Science Association, ACSA) and in the USA and globally (Citizen Science Association, CSA), as well as budding initiatives in Africa and Asia. The European Citizen Science Association has defined Ten Principles of Citizen Science (see Kelly et al., 2020):

- *Principle #1*: citizen science projects actively involve citizens in scientific endeavour that generates new knowledge or understanding
- *Principle #2:* citizen science projects have a genuine science outcome
- *Principle #3*: citizen science provides benefits to both science and society
- *Principle #4*: citizen scientists may participate in various stages of the scientific process
- *Principle #5:* citizen scientists receive feedback from the project
- *Principle #6:* citizen science, as with all forms of scientific inquiry, has limitations and biases that should be considered and controlled for
- *Principle #7*: project data/metadata are publicly available and results are published open access
- *Principle #8:* citizen scientists are suitably acknowledged by projects
- *Principle #9:* citizen science programmes offer a range of benefits and outcomes which should be acknowledged and considered in project evaluation
- *Principle #10:* leaders of projects address legal and ethical considerations of the project

Marine citizen science projects fall primarily into 5 categories:

- Action-orientated projects that encourage participation in local issues, for example collecting and categorising marine debris (e.g. the Marine Conservation Society Beachwatch project in the U.K.; https://www.mcsuk.org/what-you-can-do/join-a-beach-clean/).
- Conservation projects that promote stewardship and management such as restoring coral reefs (e.g. the Rescue a Reef project in the U.S.A; https://sharkresearch.rsmas.miami.edu/rescue-a-reef/).
- Investigation projects that answer a scientific question including monitoring threatened or invasive species (e.g. the iSeahorse project; https://projectseahorse.org/iseahorse/).
- Virtual projects that are exclusively technology-mediated for example online photo analysis (e.g. the Monitoring Tupinambás project in Brazil; https://www.zooniverse.org/projects/larissakawabe/monitore-tupinambas/about/research).
- Education projects whereby outreach is the primary goal (e.g. the Capturing Our Coasts Project in the UK; https://earthwatch.org.uk/images//Publications/CoCoast_Brochure_Web.pdf)

Marine citizen science presents a promising avenue to enhance engagement in marine conservation around the globe. Marine citizen science is growing and expanding globally and is providing innovative avenues to not just generate new scientific data but also to increase public engagement in marine science and conservation (Dean et al., 2018; Garcia-Soto et al., 2021). In addition to contributing data to science and management, citizen science projects can connect people to the oceans and further global ocean literacy (e.g. Dean et al., 2018; Freiwald et al., 2018). Appropriately designed projects reconnect participants with nature that in turn increases their awareness of the current status of the marine realm and the threat it faces. This increased awareness may then be used to promote marine stewardship, and in many cases, participants often become advocates of marine conservation.

4. Conclusion: Integrated approaches

We are facing a major crisis in the loss of the Earth's biodiversity and this has significant and far-reaching consequences for humankind. Biodiversity discovery, establishment of baselines and subsequent monitoring of trends over time are all critical components for addressing this crisis. However, it is clear that the methods that developed for taxonomy in the 18th Century are not adequate to face the current challenge (e.g. Godfray, 2007). Traditional taxonomy is simply too slow and its products, essentially static species descriptions with type material deposited in museums or other institutions, are too inaccessible to other scientists and the public (e.g. Prathapan and Rajan, 2020). The development of new or improvement of older technologies, as outlined above, provide ways to increase the discovery rate of biodiversity in the ocean but no one method provides an answer for all situations. Each approach has advantages and applications to specific questions, but they all also have limitations. Many of the studies highlighted above have identified the complementarity between methods and approaches which can help overcome such limitations and dramatically speed up the process of species discovery, description and accessibility to a range of users. These are integrated approaches to biodiversity research and below we highlight some specific examples of where complementary suite of methods have been successful and where further development may be fruitful.

An approach having some success in biodiversity discovery on land has been the application of DNA-metabarcoding on large numbers of specimens collected from the environment with subsequent clustering of DNA-sequence data being used to focus the work of taxonomists. Examples include the study of small mayflies in the Philippines (Garces et al., 2020), weevils in Sulawesi (Riedel and Narakusumo, 2019) and phorid flies in Uganda (Srivathsan et al., 2019). This approach has been termed 'turbo taxonomy' whereby organisms are sampled, preferably subjected to non-destructive DNA extraction for barcoding, assigned to MOTUs via DNA metabarcoding and then type specimens representing clusters described using conventional taxonomic approaches but including high quality digital images of new species (Riedel et al., 2013, 2014; Riedel and Narakusumo, 2019). Using these approaches tens to hundreds of new species have been identified in a very short time (Riedel et al., 2014; Riedel and Narakusumo, 2019). It is important to note that a focus of this approach is to generate both barcodes, digital images and descriptions that can be used in the rapid diagnosis of species (Riedel et al., 2013; Vences, 2020). Whilst this approach has worked well, the limitations of DNA-metabarcoding should be considered. Further developments in meta-genomic approaches or DNA-metabarcoding using third generation sequencing technologies may be a route to improved species discovery and turbo taxonomy. The turbo-taxonomy approach might be upscaled even further by biodiversity researchers working in large teams with members specialising in different aspects of the workflow outlined above (Vences, 2020).

Whilst the possibilities of taking taxonomy onto the internet have been recognised (e.g. Godfray, 2007) the full potential of this technology is not being achieved. Cybertaxonomy should go beyond the presentation of digital species descriptions to live documents that can include updates on taxonomy and species diagnosis, species distribution and ecology. Such a system should be linked to a range of tools to enable species diagnosis using a range of approaches including morphology and DNA-based evidence (e.g. Renner, 2016; Riedel et al., 2013). This will require further development of databases and data platforms with relevance to biodiversity data so that they are integrated to communicate data rapidly (at machine speeds). We do not advocate species descriptions on the basis of sequencing data alone for marine metazoans although this may need to be contemplated for taxa where type specimens may be difficult/impossible to collect, fix and preserve. An example may be some types of gelatinous zooplankton which have very delicate bodies where a combination of in-situ three-dimensional imaging combined with DNA sampling and sequencing may provide an alternative approach to conventional taxonomy.

A cybertaxonomy system should be accessible to all and should enable the participation of both professional scientists and skilled amateur naturalists in discovery and reporting of biodiversity. Information should be widely accessible in forms which are useful to society including by users from industry, governmental institutions and the public. This also means being able to trace the provenance of data from its creators through to users enabled through appropriate licensing and use of persistent identifiers. Adoption of such approaches will require harmonizing all relevant databases to deal with the large volumes of new data that will be generated, and to enable more efficient analyses for studies such as modelling changes of species distribution with climate change. We also note that the combination of cybertaxonomy and the ability to take third generation sequencers into the field has major ramifications for capacity development and citizen science. Already the first instances of species being identified in the field by citizen scientists using MinION technology and subsequently described in the field are being published (e.g. Schilthuizen et al., 2019). We also note the availability of highly effective mobile phone-based applications for citizen science focusing on biodiversity. These steps are necessary for a full global democratization of biodiversity science which includes the low- and middle-income countries (Prathapan and Rajan, 2020).

Several of the technologies we have described for biodiversity discovery and research are highly complementary and lend themselves to integrated approaches. These include, for example, the combination of eDNA approaches and active or passive acoustic survey (Easson et al., 2020; Mooney et al., 2020) or combining of traditional sampling, visual census or image-based surveys with eDNA approaches. Such complementary approaches to biodiversity research are likely to provide a more holistic view of the species present at any location and so should be encouraged and further developed.

The application of Deep Learning approaches to a range of aspects of biodiversity research including habitat definition from multibeam data, image analysis from underwater video, species identification from active acoustics or the use of passive acoustic data for biodiversity monitoring is a notable trend. Deep Learning approaches such as the use of Artificial Neural Networks are highly useful for pattern recognition and their use is likely to develop further. What has not been tested as yet is the combination of data sources for such Deep Learning approaches. For example, is it possible to combine data from image-based survey with that from passive acoustic monitoring for improved biodiversity monitoring? Another technology that will improve the spatial and temporal resolution of biodiversity sampling will be the use of autonomous platforms. These already deploy a range of sensors for physical oceanography but are increasingly capable of undertaken biological observations and sampling.

Finally, meeting the challenge of the biodiversity crisis in the ocean will not be possible without a concerted global effort. Many taxonomic institutions, taxonomists and biodiversity scientists are still based in high-income countries, a legacy of recent colonial history. To accelerate the discovery of marine species and enable all countries to meet their commitments to international treaties on biodiversity, there is pressing work that needs to be done in both capacity development and technology transfer to low- and middle-income countries. After all, it is in the coastal waters of many of these countries, especially in the tropics, where high levels of species richness and habitat diversity exist. Biodiversity loss is a global problem that requires solutions at regional and local levels.

Acknowledgements

The authors would like to acknowledge support for this paper from the High-Level Panel for a Sustainable Ocean Economy (https://oceanpanel.org/), The Nekton Foundation (https://nektonmission.org/) and The Nippon Foundation (https://www.nippon-foundation.or.jp/en). J. Assis acknowledges funding from the Foundation for Science and Technology (FCT) of Portugal via UIDB/04326/2020, UIDP/04326/2020, LA/P/0101/2020, PTDC/BIA-CBI/6515/2020 and the Scientific Employment Stimulus 2022/00861/CEECIND. A.D. Rogers and E. Ramirez-Llodra thank Rev Ocean for support throughout this project. D.P. Tittensor acknowledges support from The Jarislowsky Foundation and NSERC. E. Velarde would like to acknowledge funding from the joint fund of Fondo Mexicano para la Conservación de la Naturaleza, Lindblad Expeditions, National Geographic and the Packard Foundation. L. Woodall acknowledges support from the University of Oxford during this project.

References

Acinas, S.G., Sánchez, P., Salazar, G., Cornejo-Castillo, F.M., Sebastián, M., et al., 2021. Deep ocean metagenomes provide insight into the metabolic architecture of bathypelagic microbial communities. Comm. Biol. 4, 1–15.

Adamowicz, S.J., 2015. International barcode of life: evolution of a global research community. Genome 58, 151–162.

Agapow, P.M., Bininda-Emonds, O.R.P., Crandall, K.A., et al., 2004. The impact of species concept on biodiversity studies. Q. Rev. Biol. 79, 161–179.

Alam, I., Kamau, A.A., Ngugi, D.K., Gojobori, T., Duarte, C.M., Bajic, V.B., 2021. KAUST metagenomic analysis platform (KMAP), enabling access to massive analytics of re-annotated metagenomic data. Sci. Rep. 11, 1–14.

Alberti, A., Poulain, J., Engelen, S., et al., 2017. Viral to metazoan marine plankton nucleotide sequences from the Tara Oceans expedition. Sci. Data 4, 170093.

Alder, J., 2003. Putting the coast in the sea around us project. Sea Around US Newsletter 15, 1–2. http://seaaroundus.org/newsletter/Issue15.pdf.

Álvarez, E., Moyano, M., López-Urrutia, A., et al., 2014. Routine determination of plankton community composition and size structure: a comparison between FlowCAM and light microscopy. J. Plankton Res. 36, 170–184.

Ames, C.L., Ohdera, A.H., Colston, S.M., et al., 2021. Fieldable environmental DNA sequencing to assess jellyfish biodiversity in nearshore waters of the Florida keys, United States. Front. Mar. Sci. 8, 640527.

Amon, D.J., Gollner, S., Morato, T., Smith, C.R., Chong Chen, C., et al., 2022. Assessment of scientific gaps related to the effective environmental management of deep-seabed mining. Mar. Policy 138, 105006.

Andradi-Brown, D.A., Macaya-Solis, C., Gress, E., et al., 2016. Assessing Caribbean shallow and mesophotic reef fish communities using baited-remote underwater video (BRUV) and diver-operated video (DOV) survey techniques. PLoS One 11, e0168235.

Anton, V., Germishuys, J., Bergström, P., et al., 2021. An open-source, citizen science and machine learning approach to analyse subsea movies. Biodivers. Data J. 9, e60548.

Appeltans, W., Ahjong, S.T., Anderson, G., et al., 2012. The magnitude of global marine species diversity. Curr. Biol. 22, 2189–2202.

Ardura, A., Zaiko, A., Martinez, J.L., et al., 2015. eDNA and specific primers for early detection of invasive species—a case study on the bivalve *Rangia cuneata*, currently spreading in Europe. Mar. Environ. Res. 112, 48–55.

Asch, R.G., Cheung, W., Reygondeau, G., 2018. Future marine ecosystem drivers, biodiversity, and fisheries maximum catch potential in pacific island countries and territories under climate change. Mar. Policy 88, 285–294.

Atienza, S., Guardiola, M., Præbel, K., et al., 2020. DNA metabarcoding of deep-sea sediment communities using COI: community assessment, spatio-temporal patterns and comparison with 18S rDNA. Diversity 12, 123.

Aylagas, E., Borja, A., Irigoien, X., Rodriguez-Ezpeleta, N., 2016. Benchmarking DNA metabarcoding for biodiversity-based monitoring and assessment. Front. Mar. Sci. 3, 96.

Aylagas, E., Borja, A., Muxika, I., Rodríguez-Ezpeleta, N., 2018. Adapting metabarcoding-based benthic biomonitoring into routine marine ecological status assessment networks. Ecol. Indic. 95, 194–202.

Aylagas, E., Borja, A., Pochon, X., et al., 2020. Translational molecular ecology in practice: linking DNA-based methods to actionable marine environmental management. Sci. Total Environ. 744, 140780.

Baker, E.K., Harris, P.T., 2012. Habitat mapping and marine management. In: Harris, P.T., Baker, E.K. (Eds.), Seafloor Geomorphology as Benthic Habitat: GEOHAB Atlas of Seafloor Geomorphic Features and Benthic Habitats. Elsevier, London, UK, pp. 23–38.

Baldini, R.M., Cota-Sánchez, J.H., Aedo, C., 2021. Is the demise of plant taxonomy in sight? Maybe yes, maybe no… *Webbia*. J. Plant Taxon. Geogr. 76, 3–10.

Balls, R., 1948. Herring fishing with the echometer. J. Conseil 15, 193–206.

Balls, R., 1951. Environmental changes in herring behaviour: a theory of light avoidance as suggested by echo-sounding observations in the North Sea. J. Conseil 17, 274–298.

Bar-On, Y.M., Philips, R., Milo, R., 2018. The biomass distribution on Earth. Proc. Natl. Acad. Sci. U. S. A. 115, 6506–6511.

Barrett, R.D.H., Hebert, P.D.H., 2005. Identifying spiders through DNA barcodes. Can. J. Zool. 83, 481–491.

Basedow, S.L., Tande, K.S., Norrbin, M.F., Kristiansen, S.A., 2013. Capturing quantitative zooplankton information in the sea: performance test of laser optical plankton counter and video plankton recorder in a *Calanus finmarchicus* dominated summer situation. Prog. Oceanogr. 108, 72–80.

Baum, J.K., Worm, B., 2009. Cascading top-down effects of changing oceanic predator abundances. J. Anim. Ecol. 78, 699–714.

Bawa, K.S., Nawn, N., Chellam, R., et al., 2020. Envisioning a biodiversity science for sustaining human well-being. Proc. Natl. Acad. Sci. U. S. A. 117, 25951–25955.

Beaulieu, S.E., Szafranski, K., 2018. InterRidge Global Database of Active Submarine Hydrothermal Vent Fields. Version 3.4. http://vents-data.interridge.org.

Beck, M.W., Brumbaugh, R.D., Airoldi, L., Carranza, A., Coen, L.D., 2011. Oyster reefs at risk and recommendations for conservation, restoration, and management. Bioscience 61, 107–116.

Bennett, S., Wernberg, T., Connell, S.D., et al., 2016. The 'Great Southern Reef': social, ecological and economic value of Australia's neglected kelp forests. Mar. Freshw. Res. 67, 47–56.

Benoit-Bird, K.J., Lawson, G.L., 2016. Ecological insights from pelagic habitats acquired using active acoustic techniques. Ann. Rev. Mar. Sci. 8, 463–490.

Benoit-Bird, K.J., Waluk, C.M., 2020. Exploring the promise of broadband fisheries echosounders for species discrimination with quantitative assessment of data processing effects. J. Acoust. Soc. Am. 147, 411–427.

Bernard, A.T.F., Götz, A., Kerwath, S.E., Wilke, C.G., 2013. Observer bias and detection probability in underwater visual census of fish assemblages measured with independent double-observers. J. Exp. Mar. Biol. Ecol. 443, 75–84.

Berque, J., Matsuda, O., 2013. Coastal biodiversity management in Japanese satoumi. Mar. Policy 39, 191–200.

Beyan, C., Browman, H.I., 2020. Setting the stage for the machine intelligence era in marine science. ICES J. Mar. Sci. 77, 1267–1273.

Bicknell, A.W.J., Godley, B.J., Sheehan, E.V., et al., 2016. Camera technology for monitoring marine biodiversity and human impact. Front. Ecol. Environ. 14, 424–432.

Bilewitch, J.P., Degnan, S.M., 2011. A unique horizontal gene transfer event has provided the octocoral mitochondrial genome with an active mismatch repair gene that has potential for an unusual self-contained function. BMC Evol. Biol. 11, 228.

Blasiak, R., Wynberg, R., Grorud-Colvert, K., et al., 2020. Blue paper 4: The Ocean Genome: Conservation and the Fair, Equitable and Sustainable Use of Marine Genetic Resources. Report prepared for the (Norwegian) Prime Minister's High-Level Panel on a Sustainable Ocean Economy. 83pp. https://www.oceanpanel.org/blue-papers/ocean-genome-conservation-and-fair-equitable-and-sustainable-use-marine-genetic.

Bleidorn, C., 2016. Third generation sequencing: technology and its potential impact on evolutionary biodiversity research. Syst. Biodivers. 14, 1–8.

Boistel, R., Swoger, J., Kržič, U., et al., 2011. The future of three-dimensional microscopic imaging in marine biology. Mar. Ecol. 32, 438–452.

Bouchet, P., Bary, S., Heros, V., Marani, G., 2016. How many species of molluscs are there in the world's oceans, and who is going to describe them? In: Heros, V., Strong, E., Bouchet, P. (Eds.), Tropical Deep-Sea Benthos 29. Museum National d'Histoire Naturelle, Paris, ISBN: 978-2-85653-774-9, pp. 9–24 (Memoir Mus. Natl. Hist., 208).

Boussarie, G., Bakker, J., Wangensteen, O.S., et al., 2018. Environmental DNA illuminates the dark diversity of sharks. Sci. Adv. 4, eaap9661.

Boyce, D.G., Lotze, H.K., Tittensor, D.P., Carozza, D.A. Worm, B. 2020. Future ocean biomass losses may widen socioeconomic equity gaps. Nat. Commun. 11, 2235.

Boyce, D.G., Tittensor, D.P., Garilao, C., Henson, S., Kaschner, K., et al., 2022. A climate risk index for marine life. Nat. Clim. Change 12, 854–862.

Brack, A., Horneck, G., Wynn-Williams, D., 2001. Exo/Astrobiology in Europe. Origins Life Evol. B. 31, 459–480.

Brandt, M., 2020. Metabarcoding in the Abyss: Uncovering Deep-Sea Biodiversity Through Environmental DNA. PhD Thesis. Uni. Montpellier/Ifremer. 278 pp.

Brandt, M., Trouche, B., Quintric, L., et al., 2021. Bioinformatic pipelines combining denoising and clustering tools allow for more comprehensive prokaryotic and eukaryotic metabarcoding. Mol. Ecol. Resour. 21, 1904–1921.

Brautaset, O., Waldeland, A.U., Johnsen, E., et al., 2020. Acoustic classification in multifrequency echosounder data using deep convolutional neural networks. ICES J. Mar. Sci. 77, 1391–1400.

Britz, R., Hundsdörfer, A., Fritz, U., 2020. Funding, training, permits—the three big challenges of taxonomy. Megataxa 1, 49–52.
Broman, E., Bonaglia, S., Norkko, A., et al., 2020. High throughput shotgun sequencing of eRNA reveals taxonomic and derived functional shifts across a benthic productivity gradient. Mol. Ecol. 30, 3023–3039.
Broughton, E.A., Lough, R.G., 2006. A direct comparison of MOCNESS and video plankton recorder zooplankton abundance estimates: possible application for augmenting net sampling with video systems. Deep-Sea Res. II 53, 2789–2807.
Brown, C.J., Blondel, P., 2009. Developments in the application of multibeam sonar backscatter for seafloor habitat mapping. Appl. Acoust. 70, 1242–1247.
Brown, C.J., Beaudoin, J., Brissette, M., Gazzola, V., 2019. Multispectral multibeam echo sounder backscatter as a tool for improved seafloor characterization. Geosciences 9, 126.
Brusca, R.C., Brusca, G.J., 1990. Invertebrates. Sinauer Associates, Sunderland, MA, USA. 922 pp.
Calvert, J., Strong, J.A., Service, M., et al., 2015. An evaluation of supervised and unsupervised classification techniques for marine benthic habitat mapping using multibeam echosounder data. ICES J. Mar. Sci. 72, 1498–1513.
Campbell, R.W., Roberts, P.L., Jaffe, J., 2020. The Prince William Sound Plankton Camera: a profiling in situ observatory of plankton and particulates. ICES J. Mar. Sci. 77, 1440–1455.
Cantino, P.D., de Queiroz, K., 2020. The International Code of Phylogenetic Nomenclature (PhyloCode). CRC Press of the Taylor & Francis Group, Boca Raton, FL, USA. 149 pp.
Carradec, Q., Pelletier, E., Da Silva, C., et al., 2018. A global ocean atlas of eukaryotic genes. Nat. Comm. 9, 373.
Carradec, Q., Poulain, J., Boissin, E., et al., 2020. A framework for in situ molecular characterization of coral holobionts using nanopore sequencing. Sci. Rep. UK 10, 15893.
Carvalho, S., Aylagas, E., Villalobos, R., et al., 2019. Beyond the visual: using metabarcoding to characterize the hidden reef cryptobiome. Proc. R. Soc. B 286, 20182697.
Chang, J.J.M., Ip, Y.C.A., Bauman, A.G., Huang, D., 2020. MinION-in-ARMS: nanopore sequencing to expedite barcoding of specimen-rich macrofaunal samples from autonomous reef monitoring structures. Front. Mar. Sci. 7, 448.
Chaudhary, C., Saeedi, H., Costello, M.J., 2016. Bimodality of latitudinal gradients in marine species richness. Trends Ecol. Evol. 31, 670–676.
Clark, M.R., Consalvey, M., Rowden, A.A., 2016. Biological Sampling in the Deep Sea. Wiley Blackwell, Chichester, UK. 451pp.
Cleguer, C., Kelly, N., Tyne, J., et al., 2021. A novel method for using small unoccupied aerial vehicles to survey wildlife species and model their density distribution. Front. Mar. Sci. 8, 462.
Cochrane, S.K.J., Andersen, J.H., Berg, T., et al., 2016. What is marine biodiversity? Towards common concepts and their implications for assessing biodiversity status. Front. Mar. Sci. 3, 248.
COL, 2018. Catalogue of Life Plus: Innovating the Catalogue of Life Systems. Information Document Available at. https://github.com/CatalogueOfLife/general/blob/master/docs/CoL+slide-doc.pdf. 20 pp.
Collen, B., Turvey, S.T., Waterman, C., et al., 2011. Investing in evolutionary history: implementing a phylogenetic approach for mammal conservation. Phil. Trans. Roy. Soc. B 366, 2611–2622.
Collins, R.A., Cruickshank, R.H., 2013. The seven deadly sins of DNA barcoding. Mol. Ecol. Resour. 13, 969–975.
Convention on Biological Diversity, (1992). United Nations, New York, USA, 28pp.
Cooper, C.B., Hawn, C.L., Larson, L.R., Parrish, J.K., Bowser, G., et al., 2021. Inclusion in citizen science: the conundrum of rebranding. Science 372, 1386–1388.

Cordier, T., Angeles, I.B., Henry, N., et al., 2022. Patterns of eukaryotic diversity from the surface to the deep-ocean sediment. Sci. Adv. 8, eabj9309.

Cornford, R., Deinet, S., De Palma, A., Hill, S.L., McRae, L., et al., 2021. Fast, scalable, and automated identification of articles for biodiversity and macroecological datasets. Glob. Ecol. Biogeogr. 30, 339–347.

Costa, B., 2019. Multispectral acoustic backscatter: how useful is it for marine habitat mapping and management? J. Coast. Manag. 35, 1062–1079.

Costanza, R., D'Arge, R., de Groot, R., et al., 1997. The value of the world's ecosystem services and natural capital. Nature 387, 253–260.

Costanza, R., de Groot, R., Sutton, P., et al., 2014. Changes in the global value of ecosystem services. Glob. Environ. Chang. 26, 152–158.

Costello, M.J., 2016. Parasite rates of discovery, global species richness and host specificity. Integr. Comp. Biol. 56, 588–599.

Costello, M.J., Chaudhary, C., 2017. Marine biodiversity, biogeography, deep-sea gradients, and conservation. Curr. Biol. 27, R511–R527.

Costello, M.J., Coll, M., Danovaro, R., et al., 2010. A census of marine biodiversity knowledge, resources, and future challenges. PLoS One 5, e12110.

Costello, M.J., Wilson, S., Houlding, B., 2012. Predicting total global species richness using rates of species description and estimates of taxonomic effort. Syst. Biol. 61, 871–883.

Cowart, D.A., Pinheiro, M., Mouchel, O., et al., 2015. Metabarcoding is powerful yet still blind: a comparative analysis of morphological and molecular surveys of seagrass communities. PLoS One 10, e0117562.

Cowart, D.A., Murphy, K.R., Cheng, C.-H.C., 2018. Metagenomic sequencing of environmental DNA reveals marine faunal assemblages from the West Antarctic Peninsula. Mar. Genom. 37, 148–160.

Cowart, D.A., Matabos, M., Brandt, M.I., et al., 2020. Exploring environmental DNA (eDNA) to assess biodiversity of hard substratum faunal communities on the Lucky Strike Vent Field (Mid-Atlantic Ridge) and investigate recolonization dynamics after an induced disturbance. Front. Mar. Sci. 6, 783.

Cowie, R.H., Bouchet, P., Fontaine, B., 2022. The sixth mass extinction: fact, fiction or speculation? Biol. Rev. 97, 640–663.

Crame, J.A., Clarke, A., 1997. The historical component of marine taxonomic diversity gradients. In: Ormond, R.F.G., Gage, J.D., Angel, M.V. (Eds.), Marine Biodiversity: Patterns and Processes. Cambridge University Press, Cambridge, UK, pp. 258–273.

Creer, S., Deiner, K., Frey, S., et al., 2016. The ecologist's field guide to sequence-based identification of biodiversity. Methods Ecol. Evol. 7, 1008–1018.

Cripps, A., 2014. Sild: herring fisheries in the medieval Baltic. Blog. 3 pp https://aaroncrippsblog.wordpress.com/2014/03/26/herring/.

Cury, P., Bakun, A., Crawford, R.J.M., Jarre, A., Quiñones, R.A., et al., 2000. Small pelagics in upwelling systems: patterns of interaction and structural changes in "wasp-waist" ecosystems—ICES. J. Mar. Sci. 57, 603–618.

Danovaro, R., Dell'Anno, A., Pusceddu, A., et al., 2010. The first metazoa living in permanently anoxic conditions. BMC Biol. 8, 30.

Dasgupta, P., 2021. The Economics of Biodiversity: The Dasgupta Review. HM Treasury, London, UK. 602 pp.

Davies, R.G., Irlich, U.M., Chown, S.L., Gaston, K.J., 2010. Ambient, productive and wind energy, and ocean extent predict global species richness of procellariiform seabirds. Glob. Ecol. Biogeogr. 19, 98–110.

Davies, J.S., Stewart, H.A., Narayanaswamy, B.E., et al., 2015. Benthic assemblages of the Anton Dohrn Seamount (NE Atlantic): defining deep-sea biotopes to support habitat mapping and management efforts with a focus on vulnerable marine ecosystems. PLoS One 10, e0124815.

Davis, C.S., Thwaites, F.T., Gallager, S.M., Hu, Q., 2005. A three-axis fast-tow digital video plankton recorder for rapid surveys of plankton taxa and hydrography. Limnol. Oceanogr. Methods 3, 59–74.

Davis, G.E., Baumgartner, M.F., Bonnell, J.M., et al., 2017. Long-term passive acoustic recordings track the changing distribution of North Atlantic right whales (*Eubalaena glacialis*). from 2004 to 2014, Sci. Rep. UK 7, 13460.

De Moustier, C., 1986. Beyond bathymetry: mapping acoustic backscattering from the deep seafloor with Sea Beam. J. Acoust. Soc. Am. 79, 316–331.

De Queiroz, K., 2007. Species concepts and species delimitation. Syst. Biol. 56, 879–886.

De Queiroz, K., Gauthier, J., 1990. Phylogeny as a central principle in taxonomy: phylogenetic definitions of taxon names. Syst. Zool. 39, 307–322.

De Queiroz, K., Gauthier, J., 1994. Toward a phylogenetic system of biological nomenclature. Trends Ecol. Evol. 9, 27–31.

De Vargas, C., Audic, S., Henry, N., et al., 2015. Eukaryotic plankton diversity in the sunlit ocean. Science 348, 1261605.

Dean, A.J., Church, E.K., Loder, J., et al., 2018. How do marine and coastal citizen science experiences foster environmental engagement? J. Environ. Manage. 213, 409–416.

Delaney, D.G., Sperling, C.D., Adams, C.S., Leung, B., 2008. Marine invasive species: validation of citizen science and implications for national monitoring networks. Biol. Invasions 10, 117–128.

DeSalle, R., Egan, M.G., Siddall, M., 2005. The unholy trinity: taxonomy, species delimitation and DNA barcoding. Phil. Trans. R. Soc. B 360, 1905–1916.

Díaz, S., Pascual, U., Stenseke, M., Martín-López, B., Watson, R.T., et al., 2018. Assessing nature's contributions to people. Science 359, 270–272.

Dickinson, J.L., Bonney, R. (Eds.), 2012. Citizen science: public participation in environmental research. Comstock Publishing Associates, Cornell University Press, Ithaca, NY, p. 304pp.

Dosemagen, S., Parker, A., 2019. Citizen science across a spectrum: building partnerships to broaden the impact of citizen science. Sci. Tech. Stud. 32, 24–33.

Duarte, C.M., 2015. Seafaring in the 21st century: the *Malaspina* 2010 Circumnavigation expedition. Limnol. Oceanogr. Bull. 24, 11–14.

Duarte, C.M., Losada, I.J., Hendriks, I.E., Mazarrasa, I., Marbà, N., 2013. The role of coastal plant communities for climate change mitigation and adaptation. Nat. Clim. Change 3, 961–968.

Duarte, C.M., Agusti, S., Barbier, E., Britten, G.L., Castilla, J.C., et al., 2020a. Rebuilding marine life. Nature 580, 39–51.

Duarte, C.M., Ngugi, D.K., Alam, I., Pearman, J., Kamau, A., et al., 2020b. Sequencing effort dictates gene discovery in marine microbial metagenomes. Environ. Microbiol. 22, 4589–4603.

Dulvy, N.K., Sadovy, Y., Reynolds, J.D., 2003. Extinction vulnerability in marine populations. Fish Fish. 4, 25–64.

Dupérré, N., 2020. Old and new challenges in taxonomy: what are taxonomists up against? Megataxa 1, 59–62.

Easson, C.G., Boswell, K.M., Tucker, N., et al., 2020. Combined eDNA and acoustic analysis reflects diel vertical migration of mixed consortia in the Gulf of Mexico. Front. Mar. Sci. 7, 552.

Ebach, M.C., Holdrege, C., 2005. More taxonomy, not DNA barcoding. Bioscience 55, 822–823.

Edgar, G.J., Alexander, T.J., Lefcheck, J.S., et al., 2017a. Abundance and local-scale processes contribute to multi-phyla gradients in global marine diversity. Sci. Adv. 3, e1700419.

Edgar, G.J., Stuart-Smith, R.D., Cooper, A., et al., 2017b. New opportunities for conservation of handfishes (Family Brachionichthyidae) and other inconspicuous and threatened marine species through citizen science. Biol. Conserv. 208, 174–182.

Edmunds, P.J., Aronson, R.B., Swanson, D.W., et al., 1998. Photographic versus visual census techniques for the quantification of juvenile corals. Bull. Mar. Sci. 62, 937–946.

Egeter, B., Veríssimo, J., Lopes-Lima, M., et al., 2020. Speeding up the detection of invasive aquatic species using environmental DNA and nanopore sequencing. BioRxiv. https://doi.org/10.1101/2020.06.09.142521.

Elbrecht, V., Peinert, B., Leese, F., 2017. Sorting things out: assessing effects of unequal specimen biomass on DNA metabarcoding. Ecol. Evol. 7, 6918–6926.

Eldredge, N., Cracraft, J., 1980. Phylogenetic Patterns and the Evolutionary Process. Columbia University Press, New York. 349 pp.

Erlandson, J.M., Rick, T.C., 2010. Archaeology meets marine ecology: the antiquity of maritime cultures and human impacts on marine fisheries and ecosystems. Ann. Rev. Mar. Sci. 2, 231–251.

Estes, J.A., Duggins, D.O., 1995. Sea otters and kelp forests in Alaska: generality and variation in a community ecological paradigm. Ecol. Monogr. 65, 75–100.

Estes, J.A., Palmisano, J.F., 1974. Sea otters: their role in structuring nearshore communities. Science 185, 1058–1060.

Faulwetter, S., Dailianis, T., Vasileiadou, A., Arvanitidis, C., 2013a. Contrast enhancing techniques for the application of micro-CT in marine biodiversity studies. Microsc. Anal. Tomography Suppl. 2013, S4–S7.

Faulwetter, S., Vasileiadou, A., Kouratoras, M., et al., 2013b. Micro-computed tomography: introducing new dimensions to taxonomy. Zookeys 263, 1–45.

Fautin, D., Dalton, P., Incze, L.S., et al., 2010. An overview of marine biodiversity in United States waters. PLoS One 5, e11914.

Feldens, P., Schulze, I., Papenmeier, S., et al., 2018. Improved interpretation of marine sedimentary environments using multi-frequency multibeam backscatter data. Geosciences 8, 214.

Filbee-Dexter, K., Wernberg, T., Grace, S.P., et al., 2020. Marine heatwaves and the collapse of marginal North Atlantic kelp forests. Sci. Rep. UK 10, 13388.

Fišer, C., Robinson, C.T., Malard, F., 2018. Cryptic species as a window into the paradigm shift of the species concept. Mol. Ecol. 27, 613–635.

Flynn, J.M., Brown, E.A., Chain, F.J.J., et al., 2015. Toward accurate molecular identification of species in complex environmental samples: testing the performance of sequence filtering and clustering methods. Ecol. Evol. 5, 2252–2266.

Forrest, J.A.H., Bouchet, P.J., Barley, S.C., et al., 2020. True blue: temporal and spatial stability of pelagic wildlife at a submarine canyon. Ecosphere 12, e03423.

Freiwald, A., Rogers, A., Hall-Spencer, J., et al., 2017. Global Distribution of Cold-Water Corals. Version 5.0. Fifth Update to the Data set in Freiwald et al. (2004). http://data.unep-wcmc.org/data sets/3.

Freiwald, J., Meyer, R., Caselle, J.E., et al., 2018. Citizen science monitoring of marine protected areas: case studies and recommendations for integration into monitoring programs. Mar. Ecol. 39, e12470.

Gagné, T.O., Reygondeau, G., Jenkins, C.N., et al., 2020. towards a global understanding of the drivers of marine and terrestrial biodiversity. PLoS One 15, e0228065.

Gaines, S.D., Lubchenco, J., 1982. A unified approach to marine plant-herbivore interactions II. biogeographic patterns. Annu. Rev. Ecol. Syst. 13, 111–138.

Garces, J.M., Sartori, M., Freitag, H., 2020. Integrative taxonomy of the genus *Dudgeodes* Sartori, 2008 (Insecta, Ephemeroptera, Teloganodidae) from the Philippines with description of new species and supplementary descriptions of Southeast Asian species. Zoo Keys 910, 93–129.

Garcia-Soto, C., Seys, J.J.C., Zielinski, O., et al., 2021. Marine citizen science: current state in Europe and new technological developments. Front. Mar. Sci. 8, 621472.

Gedan, K.B., Silliman, B.R., Bertness, M.D., 2009. Centuries of human-driven change in salt marsh ecosystems. Ann. Rev. Mar. Sci. 1, 117–141.
Giovannoni, S.J., Britschgi, T.B., Moyer, C.L., Field, K.G., 1990. Genetic diversity in Sargasso Sea bacterioplankton. Nature 345, 60–63.
Giri, C., Ochieng, E., Tieszen, L.L., et al., 2011. Status and distribution of mangrove forests of the world using Earth observation satellite data: status and distributions of global mangroves. Glob. Ecol. Biogeogr. 20, 154–159.
Godfray, H.C.J., 2002. Challenges for taxonomy. Nature 417, 17–19.
Godfray, H.C.J., 2007. Linnaeus in the information age. Nature 446, 259–260.
Godfray, H.C.J., Knapp, S., 2004. Introduction. Phil. Trans. R. Soc. B 359, 559–569.
Gómez-Ríos, A., Tabik, S., Luengo, J., et al., 2019. Coral species identification with texture or structure images using a two-level classifier based on convolutional neural networks. Knowl. Based Syst. 184, 104891.
Gorsky, G., Ohman, M.D., Picheral, M., et al., 2010. Digital zooplankton image analysis using the ZooScan integrated system. J. Plankton Res. 32, 285–303.
Gouraguine, A., Moranta, J., Ruiz-Frau, A., et al., 2019. Citizen science in data and resource-limited areas: a tool to detect long-term ecosystem changes. PLoS One 14, e0210007.
Grady, J.M., Maitner, B.S., Winter, A.S., et al., 2019. Metabolic asymmetry and the global diversity of marine predators. Science 363, eaat4220.
Gregory, A.C., Zayed, A.A., Conceiçãco-Neto, N., et al., 2019. Marine DNA viral macro- and microdiversity from pole to pole. Cell 177, 1109–1123.
Greuter, W., Garrity, G., Hawkworth, D.L., et al. (Eds.), 2011. Draft BioCode (2011) Principles and rules regulating the naming of organisms. New draft, revised in November 2010. Bionomina 3, 26–44.
Guardiola, M., Uriz, M.J., Taberlet, P., et al., 2015. Deep-sea, deep-sequencing: metabarcoding extracellular DNA from sediments of marine canyons. PLoS One 10, e0139633.
Guardiola, M., Wangensteen, O.S., Taberlet, P., et al., 2016. Spatio-temporal monitoring of deep-sea communities using metabarcoding of sediment DNA and RNA. PeerJ 4, e2807.
Gutiérrez, E.E., Pine, R.H., 2017. Specimen collection crucial to taxonomy. Science 355, 1275.
Halpern, B.S., Frazier, M., Potapenko, J., et al., 2015. Spatial and temporal changes in cumulative human impacts on the world's ocean. Nat. Comm. 6, 7615.
Hamilton, C.A., Shockley, F.W., Simmons, R., et al., 2021. The future for a prominent taxonomy. Insect Syst. Divers. 5 (2), 1–2.
Hanson, M.B., Emmons, C.K., Ward, E.J., et al., 2013. Assessing the coastal occurrence of endangered killer whales using autonomous passive acoustic recorders. J. Acoust. Soc. Am. 134, 3486–3495.
Harris, P.T., 2012a. Biogeography, benthic ecology and habitat classification schemes. In: Harris, P.T., Baker, E.K. (Eds.), Seafloor Geomorphology as Benthic Habitat: GEOHAB Atlas of Seafloor Geomorphic Features and Benthic Habitats. Elsevier, London, UK, pp. 61–91.
Harris, P.T., 2012b. Surrogacy. In: Harris, P.T., Baker, E.K. (Eds.), Seafloor Geomorphology as Benthic Habitat: GEOHAB Atlas of Seafloor Geomorphic Features and Benthic Habitats. Elsevier, London, UK, pp. 93–108.
Harris, P.T., 2012c. Seafloor geomorphology-coast, shelf, and abyss. In: Harris, P.T., Baker, E.K. (Eds.), Seafloor Geomorphology as Benthic Habitat: GEOHAB Atlas of Seafloor Geomorphic Features and Benthic Habitats. Elsevier, London, UK, pp. 109–155.
Harris, P.T., Baker, E.K., 2012a. Why map benthic habitats? In: Harris, P.T., Baker, E.K. (Eds.), Seafloor Geomorphology as Benthic Habitat: GEOHAB Atlas of Seafloor Geomorphic Features and Benthic Habitats. Elsevier, London, UK, pp. 3–22.

Harris, P.T., Baker, E.K., 2012b. Seafloor Geomorphology as Benthic Habitat: GEOHAB Atlas of Seafloor Geomorphic Features and Benthic Habitats. Elsevier, London, UK. 900 pp.

Harris, R., Wiebe, P., Lenz, J., et al. (Eds.), 2000. ICES Zooplankton Methodology Manual. Academic Press, Cambridge, MA, USA. 684 pp.

Harris, P.T., Macmillan-Lawler, M., Rupp, J., Baker, E.K., 2014. Geomorphology of the oceans. Mar. Geol. 352, 4–24.

Hartwell, A.M., Voight, J.R., Wheat, C.G., 2018. Clusters of deep-sea egg-brooding octopods associated with warm fluid discharge: an ill-fated fragment of a larger, discrete population? Deep-Sea Res. I 135, 1–8.

Hatfield, R.G., Batista, F.M., Bean, T.P., et al., 2020. The application of nanopore sequencing technology to the study of dinoflagellates: a proof of concept study for rapid sequence-based discrimination of potentially harmful algae. Front. Microbiol. 11, 844.

Hatzenbuhler, C., Kelly, J.R., Martinson, J., et al., 2017. Sensitivity and accuracy of high throughput metabarcoding methods for early detection of invasive fish species. Sci. Rep. UK 7, 46393.

Hausdorf, B., 2011. Progress toward a general species concept. Evolution 65, 923–931.

Heberling, J.M., Prather, L.A., Tonsor, S.J., 2019. The changing uses of herbarium data in an era of global change: an overview using automated content analysis. Bioscience 69, 812–822.

Hebert, P.D.N., Gregory, T.R., 2005. The promise of DNA barcoding for taxonomy. Syst. Biol. 54, 852–859.

Hebert, P.D.N., Cywinska, A., Ball, S.L., deWaard, J.R., 2003a. Biological identifications through DNA barcodes. Proc. R. Soc. Lond. B 270, 313–321.

Hebert, P.D.N., Ratnasingham, S., deWaard, J.R., 2003b. Barcoding animal life: cytochrome c oxidase subunit 1 divergences among closely related species. Proc. R. Soc. Lond. B (Suppl) 270, S96–S99.

Hebert, P.D.N., Stoeckle, M.Y., Zemlak, T.S., Francis, C.M., 2004. Identification of birds through DNA barcodes. PLoS Biol. 2, e312.

Hellberg, M.E., 2006. No variation and low synonymous substitution rates in coral mtDNA despite high nuclear variation. BMC Evol. Biol. 6, 24.

Hemminga, M.A., Duarte, C.M., 2000. Seagrass Ecology. Cambridge University Press, Cambridge, UK. 312 pp.

Herman, A.W., 1988. Simultaneous measurement of zooplankton and light attenuance with a new optical plankton counter. Cont. Shelf Res. 8, 205–221.

Herman, A.W., 1992. Design and calibration of a new optical plankton counter capable of sizing small zooplankton. Deep-Sea Res. 39, 395–415.

Herman, A.W., Beanlands, B., Phillips, E.F., 2004. The next generation of optical plankton counter: the laser-OPC. J. Plankton Res. 26, 1135–1145.

Hersey, J.B., Backus, R.H., 1962. 13. Sound scattering by marine organisms. In: Hill, M.N. (Ed.), The Sea: Ideas and Observations on Progress in the Study of the Seas. Volume 1 Physical Oceanography, Interscience Publishers, a division of John Wiley & Sons, London, pp. 498–539.

Hoffman, R.C., 2005. A brief history of aquatic resource use in medieval Europe. Helgoland Mar. Res. 59, 22–30.

Hopkinson, B.M., King, A.C., Owen, D.P., et al., 2020. Automated classification of three-dimensional reconstructions of coral reefs using convolutional neural networks. PLoS One 15, e0230671.

Hoshino, T., Doi, H., Uramoto, G.I., Wörmer, L., Adhikari, R.R., Xiao, N., et al., 2020. Global diversity of microbial communities in marine sediment. Proc. Natl. Acad. Sci. U. S. A. 117, 27587–27597.

Hou, Z., Li, S., 2018. Tethyan changes shaped aquatic diversification. Biol. Rev. 93, 874–896.
Hoving, H.-J., Christiansen, S., Fabrizius, E., et al., 2019. The pelagic in-situ observation system (PELAGIOS) to reveal biodiversity, behavior, and ecology of elusive oceanic fauna. Ocean Sci. 15, 1327–1340.
Huang, D., Meier, R., Todd, P.A., Chou, L.M., 2008. Slow mitochondrial COI sequence evolution at the base of the metazoan tree and its implications for DNA barcoding. J. Mol. Evol. 66, 167–174.
Hughes, A.C., Orr, M.C., Ma, K., et al., 2021. Sampling biases shape our view of the natural world. Ecography 44, 1–11.
Hunter, R., Markert, C., 1957. Histochemical demonstration of enzymes separated by zone electrophoresis in starch gels. Science 125, 1294.
Inaba, K., 2015. Japanese marine biological stations: preface to the special issue. Reg. Stud. Mar. Sci. 2, 154–157.
IPBES, 2019. In: Díaz, S., Settele, J., Brondízio, E.S., et al. (Eds.), Summary for Policymakers of the Global Assessment Report on Biodiversity and Ecosystem Services of the Intergovernmental Science-Policy Platform on Biodiversity and Ecosystem Services. IPBES Secretariat, Bonn, Germany. 56 pp.
IPBES, 2022. Summary for policymakers of the methodological assessment regarding the diverse conceptualization of multiple values of nature and its benefits, including biodiversity and ecosystem functions and services (assessment of the diverse values and valuation of nature). Advanced unedited version, 8th July, 2022. IPBES, Bonn, Germany, 34pp.
IPCC, 2019. IPCC special report on the ocean and cryosphere in a changing climate. In: Pörtner, H.O., Roberts, D.C., Masson-Delmotte, V., Zhai, P., Tignor, M., et al. (Eds.), International Panel on Climate Change. Switzerland, Geneva.
Irigoien, X., Klevjer, T.A., Røstad, A., Martinez, U., Boyra, G., et al., 2014. Large mesopelagic fishes biomass and trophic efficiency in the open ocean. Nat. Commun. 5, 1–10.
Isaacs, J.D., Schwartzlose, R.A., 1975. Animals of the deep seafloor. Sci. Am. 233, 84–91.
IUCN, 2017. Guidelines for Using the IUCN Red List Categories and Criteria. Version 13: IUCN, Standards and Petitions Subcommittee. Gland, Switzerland.
Ivanova, N.V., Zemlak, T.S., Hanner, R.H., Hebert, P.D.N., 2007. Universal primer cocktails for fish DNA barcoding. Mol. Ecol. Notes 7, 544–548.
Jablonski, D., Roy, K., Valentine, J.W., 2006. Out of the tropics: evolutionary dynamics of the latitudinal diversity gradient. Science 314, 102–106.
Jablonski, D., Belanger, C.L., Berke, S.K., et al., 2013. Out of the tropics, but how? fossils, bridge species, and thermal ranges in the dynamics of the marine latitudinal diversity gradient. Proc. Natl. Acad. Sci. U. S. A. 110, 10487–10494.
Jackson, J.B.C., Kirby, M.X., Berger, W.H., et al., 2001. Historical overfishing and the recent collapse of coastal ecosystems. Science 293, 629–638.
Jakobsen, H.H., Carstensen, J., 2011. FlowCAM: sizing cells and understanding the impact of size distributions on biovolume of planktonic community structure. Aquat. Microb. Ecol. 65, 75–87.
Jamieson, A., 2015. The Hadal Zone: Life in the Deepest Oceans. Cambridge University Press, Cambridge, U.K. 372 pp.
Jayathilake, D., Costello, M., 2021. Version 2 of the world map of laminarian kelp benefits from more Arctic data and makes it the largest marine biome. Biol. Conserv. 257, 109099.
Jennings, S., Kaiser, M.J., Reynolds, J.D., 2001. Marine Fisheries Ecology. Blackwell Publishing, Oxford, UK. 417 pp.
Johannes, R.E., 1981. Words of the lagoon: fishing and marine lore in the Palau district of Micronesia. University of California Press, p. 245pp.

Johnson, M.W., 1948. Sound as a tool in marine ecology, from data on biological noises and the deep scattering layer. J. Mar. Res. 7, 443–458.
Jollivet, D., Desbruyeres, D., Ladrat, C., Laubier, L., 1995. Evidence for differences in the allozyme thermostability of the deep-sea hydrothermal vent polychaetes (Alvinellidae): a possible selection by habitat. Mar. Ecol. Prog. Ser. 123, 125–136.
Jones, K.R., Klein, C.J., Halpern, B.S., et al., 2018a. The location and protection status of Earth's diminishing marine wilderness. Curr. Biol. 28, 2506–2512.
Jones, F.M., Allen, C., Arteta, C., Arthur, J., Black, C., et al., 2018b. Time-lapse imagery and volunteer classifications from the Zooniverse Penguin Watch project. Sci. Data 5, 180124.
Karsenti, E., Acinas, S.G., Bork, P., et al., 2011. A holistic approach to marine ecosystems biology. PLoS Biol. 9, e1001177.
Keith, S.A., Kerswell, A.P., Connolly, S.R., 2014. Global diversity of marine macroalgae: environmental conditions explain less variation in the tropics. Glob. Ecol. Biogeogr. 23, 517–529.
Kelly, R., Fleming, A., Pecl, G.T., et al., 2020. Citizen science and marine conservation: a global review. Phil. Trans. R. Soc. Lond. B 375, 20190461.
Kerr, T., Clark, J.R., Fileman, E.S., et al., 2020. Collaborative deep learning models to handle class imbalance in FlowCam plankton imagery. IEEE Access 8, 170013–170031.
Kerswell, A.P., 2006. Global biodiversity patterns of benthic marine algae. Ecology 87, 2479–2488.
Kimura, K., 1929. On the detection of fish groups by an acoustic method. J. Imp. Fish Inst. Tokyo 24, 451–458.
Kirby, M.X., 2004. Fishing down the coast: historical expansion and collapse of oyster fisheries along continental margins. Proc. Natl. Acad. Sci. U. S. A. 101, 13096–13099.
Korneliussen, R.J. (Ed.), 2018. Acoustic target classification. ICES Cooperative Res. Rep. No. 344. 104 pp http://doi.org/10.17895/ices.pub.4567.
Korshunova, T., Picton, B., Furfaro, G., et al., 2019. Multilevel fine-scale diversity challenges the 'cryptic species' concept. Sci. Rep. UK 9, 6732.
Krämer, D., Schmidt, C., Podsiadlowski, L., et al., 2017. Unravelling the *Lineus ruber/viridis* species complex (Nemertea, Heteronemertea). Zool. Scripta 46, 111–126.
Krause-Jensen, D., Archambault, P., Assis, J., Bartsch, I., Bischof, K., et al., 2020. Imprint of climate change on pan-Arctic marine vegetation. Front. Mar. Sci. 7, 617324.
Krehenwinkel, H., Pomerantz, A., Prost, S., 2019a. Genetic biomonitoring and biodiversity assessment using portable sequencing technologies: current uses and future directions. Gene 10, 858.
Krehenwinkel, H., Pomerantz, A., Henderson, J.B., et al., 2019b. Nanopore sequencing of long ribosomal DNA amplicons enables portable and simple biodiversity assessments with high phylogenetic resolution across broad taxonomic scale. GigaScience 8, 1–16.
Krumhansl, K.A., Okamoto, D.K., Rassweiler, A., et al., 2016. Global patterns of kelp forest change over the past half-century. Proc. Natl. Acad. Sci. U. S. A. 113, 13785–13790.
Kursar, T.A., 2011. What are the implications of the Nagoya Protocol for research on biodiversity? Bioscience 61, 256–257.
Ladau, J., Sharpton, T.J., Finucane, M.M., Jospin, G., Kembel, S.W., O'dwyer, J., Koeppel, A.F., Green, J.L., Pollard, K.S., 2013. Global marine bacterial diversity peaks at high latitudes in winter. ISME J. 7, 1669–1677.
Langenkämper, D., Simon-Lledó, E., Hosking, B., et al., 2019. On the impact of citizen science-derived data quality on deep learning-based classification in marine images. PLoS One 14, e0218086.
Laroche, O., Kersten, O., Smith, C.R., Goetze, E., 2020. Environmental DNA surveys detect distinct metazoan communities across abyssal plains and seamounts in the western Clarion Clipperton Zone. Mol. Ecol. 29, 4588–4604.

Lauro, F.M., Senstius, S.J., Cullen, J., Neches, R., Jensen, R.M., et al., 2014. The common oceanographer: crowdsourcing the collection of oceanographic data. PLoS Biol. 12, e1001947.
Laverick, J.H., Andradi-Brown, D.A., Rogers, A.D., 2017. Using light-dependent scleractinia to define the upper boundary of mesophotic coral ecosystems on the reefs of Utila, Honduras. PLoS One 12, e0183075.
Le Guillarme, N., Thuiller, W., 2022. TaxoNERD: deep neural models for the recognition of taxonomic entities in the ecological and evolutionary literature. Methods Ecol. Evol. 13, 625–641.
Lejzerowicz, F., Esling, P., Pillet, L., et al., 2015. High-throughput sequencing and morphology perform equally well for benthic monitoring of marine ecosystems. Sci. Rep. UK 5, 13932.
Lejzerowicz, F., Gooday, A.J., Angeles, I.B., et al., 2021. Eukaryotic biodiversity and spatial patterns in the Clarion-Clipperton Zone and other abyssal regions: insights from sediment DNA and RNA metabarcoding. Front. Mar. Sci. 8, 671033.
Leray, M., Knowlton, N., 2017. Random sampling causes the low reproducibility of rare eukaryotic OTUs in Illumina COI metabarcoding. PeerJ 5, e3006.
Li, Q., Sun, X., Dong, J., et al., 2020. Developing a microscopic image dataset in support of intelligent phytoplankton detection using deep learning. ICES J. Mar. Sci. 77, 1427–1439.
Liebenberg, L., Lombard, M., Shermer, M., Xhukwe, U., Biesele, M., et al., 2021. Tracking science: an alternative for those excluded by citizen science. Citiz. Sci. Theory Prac. 6, 6.
Lombard, F., Boss, E., Waite, A.M., et al., 2019. Globally consistent quantitative observations of planktonic ecosystems. Front. Mar. Sci. 6, 196.
Luczkovich, J.J., Mann, D.A., Rountree, R.A., 2008. Passive acoustics as a tool in fisheries science. Trans. Am. Fish. Soc. 137, 533–541.
Mächler, E., Walser, J.C., Altermatt, F., 2021. Decision-making and best practices for taxonomy-free environmental DNA metabarcoding in biomonitoring using Hill numbers. Mol. Ecol. 30, 3326–3339.
Maestri, S., Cosentino, E., Paterno, M., et al., 2019. A rapid and accurate MinION-based workflow for tracking species biodiversity in the field. Gene 10, 468.
Manktelow, M., 2010. History of Taxonomy. Lecture, Uppsala University, Uppsala, Sweden. 8 pp http://www.atbi.eu/summerschool/files/summerschool/Manktelow_Syllabus.pdf.
Marques, T.A., Thomas, L., Ward, J., et al., 2009. Estimating cetacean population density using fixed passive acoustic sensors: an example with Blainville's beaked whales. J. Acoust. Soc. Am. 125, 1982–1994.
Marti, M., 1922. Brevet d'Invention XII, 3 No. 568-075, Oscillographe enregistreur continu à grande sensibilité et son application au sondage continu par le son. (7 Sept. 1922).
Martin, S.L., Ince, L., Grooves, T., 2016. An ecosystem services perspective for the oceanic eastern tropical pacific: commercial fisheries, carbon storage, recreational fishing, and biodiversity. Front. Mar. Sci. 3, 50.
Masood, E., 2018. Battle over biodiversity: an ideological clash could undermine a crucial assessment of the world's disappearing plant and animal life. Nature 560, 423–425.
May, R.M., 1988. How many species are there on Earth? Science 241, 1441–1449.
May, R.M., 1994. Biological diversity: differences between land and sea. Phil. Trans. R. Soc. Lond. B 343, 105–111.
Mayr, E., 1942. Systematics and the Origin of Species. Columbia University Press, NY, USA. 334 pp.
Mayr, E., 1963. Animal Species and Evolution. Belknap Press, Cambridge, MA, USA, p. 797.
McCauley, D.J., Pinsky, M.L., Palumbi, S.R., et al., 2015. Marine defaunation: animal loss in the global ocean. Science 347, 1255641.

McClenachan, L., Cooper, A.B., 2008. Extinction rate, historical population structure and ecological role of the Caribbean monk seal. Proc. R. Soc. Lond. B 275, 1351–1358.

Mcfadden, C.S., Benayahu, Y., Pante, E., et al., 2011. Limitations of mitochondrial gene barcoding in Octocorallia. Mol. Ecol. Res. 11, 19–31.

McKenzie, L.J., Nordlund, L.M., Jones, B.L., Cullen-Unsworth, L.C., Roelfsema, C., Unsworth, R.K.F., 2020. The global distribution of seagrass meadows. Environ. Res. Lett. 15, 074041.

McOwen, C., Weatherdon, L.V., Bochove, J., et al., 2017. A global map of saltmarshes. Biodivers. Data J. 5, e11764.

Medeiros, M.C., Pinto, A.S., Rodrigues dos Santos, D., Martel, G., Lopes, S. de F., Mourao J. da S., 2022. Folk taxonomy and scientific nomenclature: working together for conservation of fishery resources in Brazil. J. Nat. Conserv. 68, 126214.

Menandro, P.S., Bastos, A.C., 2020. Seabed mapping: a brief history from meaningful words. Geosciences 10, 273.

Menegon, M., Cantaloni, C., Rodriguez-Prieto, A., et al., 2017. On site DNA barcoding by nanopore sequencing. PLoS One 12, e0184741.

Mentel, M., Martin, W., 2010. Anaerobic animals from an ancient, anoxic ecological niche. BMC Biol. 8, 32.

Meyer, C.P., Paulay, G., 2005. DNA barcoding: error rates based on comprehensive sampling. PLoS Biol. 3, e422.

Miller, E.C., Wiens, J.J., 2017. Extinction and time help drive the marine-terrestrial biodiversity gradient: is the ocean a deathtrap? Ecol. Lett. 20, 911–921.

Miloslavich, P., Webb, T., Snelgrove, P. et al., 2016. Chapter 35. Extent of Assessment of Marine Biological Diversity. In: Inniss, L., Simcock, A. (Coordinators) The First Integrated Marine Assessment: World Ocean Assessment I. United Nations General Assembly, United Nations, New York, 58 pp.

Miranda, R., Miqueleiz, I., Darwall, W., et al., 2022. Monitoring extinction risk and threats of the world's fishes based on the sampled red list index. Rev. Fish Biol. Fish. 32, 975–991.

Mooney, T.A., Iorio, L.D., Lammers, M., et al., 2020. Listening forward: approaching marine biodiversity assessments using acoustic methods. R. Soc. Open Sci. 7, 201287.

Mora, C., Tittensor, D.P., Adl, S., et al., 2011. How many species are there on Earth and in the ocean? PLoS Biol. 9, e1001127.

Morris, R.M., Rappé, M.S., Connon, S.A., et al., 2002. SAR11 clade dominates ocean surface bacterioplankton communities. Nature 420, 806–810.

Morris, K., Tyler, P.A., Murton, B., Rogers, A.D., 2012. Lower Bathyal and abyssal distribution of coral in the axial volcanic ridge if the Mid-Atlantic ridge at 45°N. Deep-Sea Res. I 62, 32–39.

Morrissey, R.P., Ward, J., DiMarzio, N., et al., 2006. Passive acoustic detection and localization of sperm whales (*Physeter macrocephalus*) in the tongue of the ocean. Appl. Acoust. 67, 1091–1105.

Moura, C.J., Harris, D.J., Cunha, M.R., Rogers, A.D., 2008. DNA barcoding reveals cryptic diversity in marine hydroids (Cnidaria, Hydrozoa) from coastal and deep-sea environments *Zool.* Scripta 37, 93–108.

Moura, C.J., Cunha, M.R., Porteiro, F.M., Rogers, A.D., 2011. The use of the DNA barcode gene 16S mRNA for the clarification of taxonomic problems within the family Sertulariidae (Cnidaria, Hydrozoa). Zool. Scripta 40, 520–537.

Murray, J., Hjort, J., 1912. The Depths of the Ocean: A General Account of the Modern Science of Oceanography Based Largely on the Scientific Researches of the Norwegian Steamer Michael Sars in the North Atlantic. Reprinted 1965 by J. Cramer. Weinheaim, Wheldon & Wesley Ltd., Codicote, Hertfordshire, U.K. 821pp + 4 Maps and 9 Plates.

Myers, R.A., Baum, J.K., Shepherd, T.D., Powers, S.P., Peterson, C.H., 2007. Cascading effects of the loss of apex predatory sharks from a coastal ocean. Science 315, 1846–1850.
Naeem, S., 2002. Ecosystem consequences of biodiversity loss: evolution of a paradigm. Ecology 83, 1537–1552.
Nayak, A.R., Malkiel, E., McFarland, M.N., et al., 2021. A review of holography in the aquatic sciences: in situ characterization of particles, plankton, and small-scale biophysical interactions. Front. Mar. Sci. 7, 572147.
Nei, M., 1972. Genetic distance between populations. Am. Nat. 106, 283–292.
Nelleman, C., Corcoran, E., Duarte, C.M., Valdes, L., DeYoung, C., et al., 2009. Blue carbon: the role of healthy oceans in binding carbon. UNEP/FAO/UNESCO/IUCN/CSIC.
Nixon, K.C., Carpenter, J.M., Stevenson, D.W., 2003. The PhyloCode is fatally flawed, and the "Linnaean" system can easily be fixed. Bot. Rev. 69, 111–120.
Norman, B.M., Holmberg, J.A., Arzoumanian, Z., et al., 2017. Undersea constellations: the global biology of an endangered marine megavertebrate further informed through citizen science. Bioscience 67, 1029–1043.
Noyer, C., Abot, A., Trouilh, L., et al., 2015. Phytochip: development of a DNA-microarray for rapid and accurate identification of *Pseudonitzschia* spp and other harmful algal species. J. Microbiol. Methods 112, 55–66.
Obiol, A., Giner, C.R., Sánchez, P., Duarte, C.M., Acinas, S.G., Massana, R., 2020. A metagenomic assessment of microbial eukaryotic diversity in the global ocean. Mol. Ecol. Resour. 20, 718–731.
Oliver, T.H., Heard, M.S., Isaac, N.J., Roy, D.B., Procter, D., et al., 2015. Biodiversity and resilience of ecosystem functions. Trends Ecol. Evol. 30, 673–684.
Olson, R.J., Sosik, H.M., 2007. A submersible imaging-in-flow instrument to analyze nano and microplankton: imaging FlowCytobot. Limnol. Oceanogr. Meth. 5, 195–203.
Ortega, A., Geraldi, N.R., Alam, I., Kamau, A.A., Acinas, S.G., et al., 2019. Important contribution of macroalgae to oceanic carbon sequestration. Nat. Geosci. 12, 748–754.
Pacoureau, N., Rigby, C.L., Kyne, P.M., et al., 2021. Half a century of global decline in oceanic sharks and rays. Nature 589, 567–571.
Pappalardo, P., Fernández, M., 2014. Mode of larval development as a key factor to explain contrasting effects of temperature on species richness across oceans. Glob. Ecol. Biogeogr. 23, 12–23.
Parks, D.H., Rinke, C., Chuvochina, M., Chaumeil, P.A., Woodcroft, B.J., et al., 2017. Recovery of nearly 8,000 metagenome-assembled genomes substantially expands the tree of life. Nat. Microbiol. 2, 1533–1542.
Parravicini, V., Morri, C., Ciribilli, G., et al., 2009. Size matters more than method: visual quadrats vs photography in measuring human impact on Mediterranean rocky reef communities. Estuar. Coast. Shelf S 81, 359–367.
Parsons, M.J.G., Lin, T.-H., Mooney, T.A., Erbe, C., Juanes, F., et al., 2022. Sounding the call for a global library of underwater biological sounds. Front. Ecol. Evol. 10, 810156.
Pawlowski, J., Kelly-Quinn, M., Altermatt, F., et al., 2018. The future of biotic indices in the ecogenomic era: integrating (e)DNA metabarcoding in biological assessment of aquatic ecosystems. Sci. Total Environ. 637–638, 1295–1310.
Pearman, J.K., Chust, G., Aylagas, E., et al., 2020. Pan-regional marine benthic cryptobiome biodiversity patterns revealed by metabarcoding autonomous reef monitoring structures. Mol. Ecol. 29, 4882–4897.
Pennisi, E., 2001. Linnaeus's Last Stand? Science 291, 2304–2307.
Pesant, S., Not, F., Picheral, M., et al., 2015. Open science resources for the discovery and analysis of Tara Oceans data. Sci. Data 2, 150023.
Piccino, P., Viard, F., Sarradin, P.-M., Le Bris, N., Le Guen, D., Jollivet, D., 2004. Thermal selection of PGM allozymes in newly founded populations of the thermotolerant vent polychaete Alvinella pompejana. Proc. R. Soc. Lond. B 271, 2351–2359.

Picciulin, M., Kéver, L., Parmentier, E., Bolgan, M., 2019. Listening to the unseen: passive acoustic monitoring reveals the presence of a cryptic fish species. Aquat. Conserv. 29, 202–210.

Pikitch, E.K., 2018. A tool for finding rare marine species. Science 360, 1180–1182.

Pillay, T., Cawthrab, H.C., Lombard, A.T., 2020. Characterisation of seafloor substrate using advanced processing of multibeam bathymetry, backscatter, and sidescan sonar in Table Bay, South Africa. Mar. Geol. 429, 106332.

Pinto, Â.P., Mejdalani, G., Mounce, R., et al., 2021. Are publications on zoological taxonomy under attack? R. Soc. Open Sci. 8, 201617.

Pomerantz, A., Peñafiel, N., Arteaga, A., et al., 2018. Real-time DNA barcoding in a rainforest using nanopore sequencing: opportunities for rapid biodiversity assessments and local capacity building. Gigascience 7, 1–14.

Pomeroy, R., Parks, J., Mrakovcich, K.L., LaMonica, C., 2016. Drivers and impacts of fisheries scarcity, competition, and conflict on maritime security. Mar. Policy 67, 94–104.

Pompa, S., Ehrlich, P., Ceballos, G., 2011. Global distribution and conservation of marine mammals. Proc. Natl. Acad. Sci. U. S. A. 108, 13600–13605.

Pont-Kingdon, G., Okada, N.A., Macfarlane, J.L., et al., 1998. Mitochondrial DNA of the coral *Sarcophyton glaucum* contains a gene for a homologue of bacterial MutS: a possible case of gene transfer from the nucleus to the mitochondrion. J. Mol. Evol. 46, 419–431.

Porter, T.M., Hajibabaei, M., 2018. Scaling up: a guide to high-throughput genomic approaches for biodiversity analysis. Mol. Ecol. 27, 313–338.

Prathapan, K.D., Rajan, P.D., 2020. Advancing taxonomy in the global south and completing the grand linnaean enterprise. Megataxa 1, 73–77.

Prato, G., Thiriet, P., Di Franco, A., Francour, P., 2017. Enhancing fish underwater visual census to move forward assessment of fish assemblages: an application in three Mediterranean Marine Protected Areas. PLoS One 12, e0178511.

Radeta, M., Nunes, N.J., Vasconcelos, D., Nisi, V., 2018. POSEIDON—Passive-acoustic ocean sensor for entertainment and interactive data-gathering in opportunistic nautical-activities. In: DIS '18: Proceedings of the 2018 Designing Interactive Systems Conference, pp. 999–1011. https://dl.acm.org/doi/10.1145/3196709.3196752.

Radulovici, A.E., Vieira, P.E., Duarte, S., et al., 2021. Revision and annotation of DNA barcode records for marine invertebrates: report of the 18th iBOL conference hackathon. BioRxiv. https://doi.org/10.1101/2021.03.07.434272. Preprint.

Ramirez-Llodra, E., Hilario, A., Paulsen, E., et al., 2020. Benthic communities on the Mohn's treasure mound: implications for management of seabed mining in the Arctic Mid-Ocean Ridge. Front. Mar. Sci. 7, 490.

Rampeletto, P.H., 2010. Resistance of microorganisms to extreme environmental conditions and its contribution to astrobiology. Sustainability 2, 1602–1623.

Ratnasingham, S., Hebert, P.D.N., 2007. BOLD: The barcode of life data system (www.barcodinglife.org). Mol. Ecol. Notes 7, 355–364.

Raven, P.H., Berlin, B., Breedlove, D.E., 1971. The origins of taxonomy. Science 174, 1210–1213.

Renner, S.S., 2016. A return to Linnaeus's focus on diagnosis, not description: the use of DNA characters in the formal naming of species. Syst. Biol. 65, 1085–1095.

Reygondeau, G., 2019. Current and future biogeography of exploited marine exploited groups under climate change. In: Cisneros-Montemayor, A.M., Cheung, W.W.L., Ota, Y. (Eds.), Predicting Future Oceans: Sustainability of Ocean and Human Systems Amidst Global Environmental Change. Elsevier, Oxford, UK, pp. 87–101.

Rice, A.L., Aldred, R.G., Billett, D.S.M., Thurston, M.H., 1979. The combined use of an epibenthic sledge and a deep-sea camera to give quantitative relevance to macro-benthos samples. Ambio Special Rep. 6, 59–72.

Richards, R.A., 2015. The Species Problem: A Philosophical Analysis. Cambridge University Press, Cambridge, UK. 236 pp.
Riedel, A., Narakusumo, R.P., 2019. One hundred and three new species of *Trigonopterus* weevils from Sulawesi. ZooKeys 828, 1–153.
Riedel, A., Sagata, K., Suhardjono, Y.R., et al., 2013. Integrative taxonomy on the fast track—towards more sustainability in biodiversity research. Front. Zool. 10, 15.
Riedel, A., Tänzler, R., Balke, M., et al., 2014. Ninety-eight new species of *Trigonopterus* weevils from Sundaland and the Lesser Sunda Islands. ZooKeys 467, 1–162.
Roberge, J.-M., Angelstam, P., 2004. Usefulness of the umbrella species concept as a conservation tool. Conserv. Biol. 18, 76–85.
Roberts, C., 2007. The Unnatural History of the Sea. Gaia (Division of Octopus Publishers), London, UK. 448 pp.
Rogers, A.D., 1994. The biology of seamounts. Adv. Mar. Biol. 30, 305–350.
Rogers, A.D., 2018. The biology of seamounts: 25 years on. Adv. Mar. Biol. 79, 137–224.
Rogers, C.M., 2019. Taxonomy and the history of science: a critical analysis of four historic publications. KSU J. Hon. College Capstones Theses 23. 39 pp https://digitalcommons.kennesaw.edu/honors_etd/23.
Rogers, A.D., Nash, R.D.M., 1996. A new species of *Ochetostoma* (Echiura, Echiuridae) found in the Azores with notes on its ecology. J. Mar. Biol. Assoc. UK 76, 467–478.
Rogers, A.D., Junoy, J., Gibson, R., Thorpe, J.P., 1993. Enzyme electrophoresis, genetic identity and description of a new genus and species of heteronemertean (Nemertea, Anopla) from northwestern Spain and North Wales. Hydrobiologia 266, 219–238.
Rogers, A., Aburto-Oropeza, O., Appeltans, W., et al., 2020. Blue Paper 10: Critical Habitats and Biodiversity: Inventory, Thresholds, and Governance. Report prepared for the (Norwegian) Prime Minister's High-Level Panel on a Sustainable Ocean Economy. 85 pp https://www.oceanpanel.org/blue-papers/critical-habitats-and-biodiversity-inventory-thresholds-and-governance.
Rogers, A.D., Baco-Taylor, A., Escobar-Briones, E., et al., 2021. Marine genetic resources in areas beyond national jurisdiction: promoting marine scientific research and enabling equitable benefit sharing. Front. Mar. Sci. 8, 667274.
Roux, S., Brum, J.R., Dutilh, B.E., Sunagawa, S., et al., 2016. Ecogenomics and potential biogeochemical impacts of globally abundant ocean viruses. Nature 537, 689–697.
Roy, K., Jablonski, D., Valentine, J.W., Rosenberg, G., 1998. marine latitudinal diversity gradients: tests of causal hypotheses. Proc. Natl. Acad. Sci. U. S. A. 95, 3699–3702.
Rozwadowski, H.M., 2005. Fathoming the Ocean: The Discovery and Exploration of the Deep Sea. The Belknap Press of Harvard University Press, Cambridge, MA, USA. 276 pp.
Sadovy, Y., Cheung, W.L., 2003. Near extinction of a highly fecund fish: the one that nearly got away. Fish Fish. 4, 86–99.
Sahlin, K., Medvedev, P., 2021. Error correction enables use of Oxford Nanopore technology for reference-free transcriptome analysis. Nat. Commun. 12, 2.
Salazar, G., Sunagawa, S., 2017. Marine microbial diversity. Curr. Biol. 27, R489–R494.
Salazar, G., Paoli, L., Alberti, A., et al., 2019. Gene expression changes and community turnover differentially shape the global ocean metatranscriptome. Cell 179, 1068–1083.
Salisbury, D.P., Clark, C.W., Rice, A.N., 2015. Right whale occurrence in the coastal waters of Virginia, U.S.A.: endangered species presence in a rapidly developing energy market. Mar. Mamm. Sci. 32, 508–519.
Salman, A., Jalal, A., Shafait, F., et al., 2016. Fish species classification in unconstrained underwater environments based on deep learning. Limnol. Oceanogr. Meth. 14, 570–585.
Salman, A., Siddiqui, S.A., Shafait, F., et al., 2019. Automatic fish detection in underwater videos by a deep neural network-based hybrid motion learning system. ICES J. Mar. Sci. 77, 1295–1307.

Samoilys, M.A., Carlos, G., 2000. Determining methods of underwater visual census for estimating the abundance of coral reef fishes. Environ. Biol. Fishes 57, 289–304.

Santos-Domínguez, D., Torres-Guijarro, S., Cardenal-López, A., Pena-Gimenez, A., 2016. ShipsEar: an underwater vessel noise database. Appl. Acoust. 113, 64–69.

Sathirathai, S., Barbier, E.B., 2001. Valuing mangrove conservation in Southern Thailand. Contemp. Econ. Policy 19, 109–122.

Saunders, T.E., 2020. Taxonomy at a crossroads: communicating value, building capability, and seizing opportunities for the future. Megataxa 1, 63–66.

Scheffers, B.R., Joppa, L.N., Pimm, S.L., Laurance, W.F., 2012. What we know and don't know about Earth's missing biodiversity. Trends Ecol. Evol. 27, 501–510.

Schevill, W.E., Backus, R.H., Hersey, J.B., 1962. In: Hill, M.N. (Ed.), The Sea: Ideas and Observations on Progress in the Study of the Seas. vol. 1 Physical Oceanography, Interscience Publishers, a division of John Wiley & Sons, London, pp. 540–566.

Schilthuizen, M., Clavera, A.P., Khoo, M.S., et al., 2019. Bringing the lab to the field: a new lowland Microparmarion semi-slug (Gastropoda: Ariophantidae) described and DNA-barcoded in the forest. J. Moll. Stud. 85, 35–40.

Schimel, A.C.G., Brown, C.J., Ierodiaconou, D., 2020. Automated filtering of multibeam water-column data to detect relative abundance of giant kelp (*Macrocystis pyrifera*). Remote Sen.-Basel 12, 1371.

Schofield, G., Esteban, N., Katselidis, K., Hays, G., 2019. Drones for research on sea turtles and other marine vertebrates—a review. Biol. Conserv. 238, 108214.

Scholin, C.A., Birch, J., Jensen, S., Marin III, R., Massion, E., Pargett, D., Preston, C., Roman, B., Ussler III, W., 2017. The quest to develop ecogenomic sensors: a 25-Year history of the Environmental Sample Processor (ESP) as a case study. Oceanography 30, 100–113.

Schweiger, O., Klotz, S., Durka, W., Kühn, I., 2008. A comparative test of phylogenetic diversity indices. Oecologia 157, 485–495.

Seeleuthner, Y., Mondy, S., Lombard, V., et al., 2018. Single-cell genomics of multiple uncultured stramenopiles reveals underestimated functional diversity across oceans. Nat. Comm. 9, 310.

Selvanes, A.G.V., Devine, J., Jensen, K.H., et al., 2017. Marine Ecological Field Methods: A Guide for Marine Biologists and Fisheries Scientists. Wiley-Blackwell, Chichester. 240 pp.

Semmouri, I., De Schamphelaere, K.A.C., Mees, J., et al., 2019. Evaluating the potential of direct RNA nanopore sequencing: metatranscriptomics highlights possible seasonal differences in a marine pelagic crustacean zooplankton community. Mar. Environ. Res. 153, 104836.

Short, F., Carruthers, T., Dennison, W., Waycott, M., 2007. Global seagrass distribution and diversity: a bioregional model. J. Exp. Mar. Biol. Ecol. (Special Issue) 350, 3–20.

Sieracki, C.K., Sieracki, M.E., Yentsch, C.S., 1998. An imaging-in-flow system for automated analysis of marine microplankton. Mar. Ecol. Prog. Ser. 168, 285–296.

Singer, G.A.C., Shekarriz, S., McCarthy, A., et al., 2020. The utility of a metagenomics approach for marine biomonitoring. BioRxiv. https://doi.org/10.1101/2020.03.16.993667.

Sosik, H.M., Olson, R.J., 2007. Automated taxonomic classification of phytoplankton sampled with imaging-in-flow cytometry. Limnol. Oceanogr. Meth. 5, 204–216.

Srivathsan, A., Hartop, E., Puniamoorthy, J., et al., 2019. Rapid, large-scale species discovery in hyperdiverse taxa using 1D MinION sequencing. BMC Biol. 17, 96.

Staaterman, E., Ogburn, M.B., Altieri, A.H., et al., 2017. Bioacoustic measurements complement visual biodiversity surveys: preliminary evidence from four shallow marine habitats. Mar. Ecol. Prog. Ser. 575, 207–215.

Staehr, P.A.U., Dahl, K., Buur, H., Göke, C., Sapkota, R., et al., 2022. Environmental DNA monitoring of biodiversity hotspots in Danish marine waters. Front. Mar. Sci. 8, 800474.

Standing, G., 2022. The Blue Commons: Rescuing the Economy of the Sea. Penguin Random House, London, UK. 592pp.
Stefanoudis, P.V., Rivers, M., Smith, S.R., et al., 2019a. Low connectivity between shallow, mesophotic and rariphotic zone benthos. R. Soc. Open Sci. 6, 190958.
Stefanoudis, P.V., Gress, E., Pitt, J.M., et al., 2019b. Depth-dependent structuring of reef fish assemblages from the shallows to the rariphotic zone. Front. Mar. Sci. 6, 307.
Stork, N.E., McBroom, J., Gely, C., Hamilton, A.J., 2015. New approaches narrow global species estimates for beetles, insects, and terrestrial arthropods. Proc. Natl. Acad. Sci. U. S. A. 112, 7519–7523.
Strickland, H.E., 1843. Report of a committee appointed "to consider of the rules by which the nomenclature of zoology may be established on a uniform and permanent basis". In: Report of the twelfth meeting of the British Association for the Advancement of Science (Manchester, 1842), pp. 105–121.
Strong, J.A., Andonegi, E., Bizsel, K.C., et al., 2015. Marine biodiversity and ecosystem function relationships: the potential for practical monitoring applications. Estuar. Coast. Shelf S 161, 46–64.
Stuart-Smith, R.D., Bates, A.E., Lefcheck, J.S., Duffy, J.E., Baker, S.C., et al., 2013. Integrating abundance and functional traits reveals new global hotspots of fish diversity. Nature 501, 539–542.
Sture, Ø., Ludvigsen, M., Scheide, M.S., Thorsnes, T., 2018. Recognition of cold-water corals in synthetic aperture sonar imagery. In: 2018 IEEE/OES Autonomous Underwater Vehicle Workshop (AUV), 2018, pp. 1–6, https://doi.org/10.1109/AUV.2018.8729718.
Sullivan, J.M., Constant, V., Lubchenco, J., 2019. Extinction threats to life in the ocean and opportunities for their amelioration. In: Dasgupta, P., Raven, P., McIvor, A. (Eds.), Biological Extinction: New Perspectives. Cambridge University Press, Cambridge, UK, pp. 113–137.
Sumaila, U.R., Lam, V.W.Y., Miller, D.D., et al., 2015. Winners and losers in a world where the high seas is closed to fishing. Sci. Rep. UK 5, 8481.
Sunagawa, S., Coelho, L.P., Chaffron, S., et al., 2015. Structure and function of the global ocean microbiome. Science 348, 1261359.
Sund, O., 1935. Echo sounding in fisheries research. Nature 135, 953.
Sward, D., Monk, J., Barrett, N., 2019. A systematic review of remotely operated vehicle surveys for visually assessing fish assemblages. Front. Mar. Sci. 6, 134.
Taberlet, P., Bonin, A., Zinger, L., Coissac, E., 2018. Environmental DNA For Biodiversity Research and Monitoring. Oxford University Press, Oxford, UK. 253 pp.
Tashiro, T., Ishida, A., Hori, M., et al., 2017. Early trace of life from 3.95 Ga sedimentary rocks in Labrador, Canada. Nature 549, 516–518.
Taylor, M.L., Yesson, C., Agnew, D.J., et al., 2013. Using fisheries by-catch data to predict octocoral habitat suitability around South Georgia. J. Biogeogr. 40, 1688–1701.
Thomas, N., Lucas, R., Bunting, P., et al., 2017. Distribution and drivers of global mangrove forest change, 1996–2010. PLoS One 12, e0179302.
Thorpe, J.P., 1982. The molecular clock hypothesis: biochemical evolution, genetic differentiation and systematics. Annu. Rev. Ecol. Syst. 13, 139–168.
Thorpe, J.P., Solé-Cava, A.M., 1994. The use of allozyme electrophoresis in invertebrate systematics. Zool. Scripta 23, 3–18.
Thurstan, R.H., Brockington, S., Roberts, C.M., 2010. The effects of 118 years of industrial fishing on UK bottom trawl fisheries. Nature Comm. 1, 15.
Tittensor, D.P., Mora, C., Jetz, W., et al., 2010. Global patterns and predictors of marine biodiversity across taxa. Nature 466, 1098–1101.
UNEP-WCMC, Short, F.T, 2018. Global Distribution of Seagrasses. Version 6.0. Sixth update to the data layer used in Green and Short (2003). UNEP-WCMC, Cambridge, UK. http://data.unep-wcmc.org/datasets/7.

UNEP-WCMC, WorldFish Centre, WRI, TNC, 2018. Global Distribution of Warm-Water Coral Reefs, Compiled from Multiple Sources Including the Millennium Coral Reef Mapping Project. Version 4.0. Includes contributions from IMaRS-USF and IRD (2005), IMaRS-USF (2005), and Spalding et al. (2001). UNEP-WCMC, Cambridge, UK. http://data.unep-wcmc.org/datasets/1.
Uthicke, S., Robson, B., Doyle, J.R., Logan, M., Pratchett, M.S., Lamare, M., 2022. Developing an effective marine eDNA monitoring: eDNA detection at pre-outbreak densities of corallivorous seastar (*Acanthaster* cf. *solaris*). Sci. Total Environ. 851, 158143.
Valentini, A., Taberlet, P., Miaud, C., et al., 2016. Next-generation monitoring of aquatic biodiversity using environmental DNA metabarcoding. Mol. Ecol. 25, 929–942.
Van Dover, C.L., 2000. The Ecology of Deep-Sea Hydrothermal Vents. Princeton University Press, Princeton, NJ, USA. 448 pp.
Van der Velde, T., Milton, D.A., Lawson, T.J., et al., 2017. Comparison of marine debris data collected by researchers and citizen scientists: is citizen science data worth the effort? Biol. Conserv. 208, 127–138.
Vandersea, M.W., Kibler, S.R., Van Sant, S.B., et al., 2017. qPCR assays for *Alexandrium fundyense* and *A. ostenfeldii* (Dinophyceae) identified from Alaskan waters and a review of species-specific *Alexandrium* molecular assays. Phycologia 56, 303–320.
Veitayaki, J., 1997. Traditional marine resource management practices used in the Pacific Islands: an agenda for change. Ocean Coast. Manag. 37, 123–136.
Vences, M., 2020. The promise of next-generation taxonomy. Megataxa 1, 35–38.
Vigoureux, P., Hersey, J.B., 1962. 12. Sound in the sea. In: Hill, M.N. (Ed.), The Sea: Ideas and Observations on Progress in the Study of the Seas. Volume 1 Physical Oceanography. Interscience Publishers, a division of John Wiley & Sons, London, pp. 476–497.
Villon, S., Mouillot, D., Chaumont, M., et al., 2018. A deep learning method for accurate and fast identification of coral reef fishes in underwater images. Eco. Inform. 48, 238–244.
Von Deimling, J.S., Brockhoff, J., Greinert, J., 2007. Flare imaging with multibeam systems: data processing for bubble detection at seeps. Geochem. Geophys. Geosyst. 8, Q06004.
Wada, T., Doi, H., Togaki, D., et al., 2020. Exploring a legendary giant squid: an environmental DNA approach. Mar. Biol. 167, 160.
Wafar, M., Venkataraman, K., Baban Ingole, B., et al., 2011. State of knowledge of coastal and marine biodiversity of Indian Ocean countries. PLoS One 6, e14613.
Wäldchen, J., Mäder, P., 2018. Machine learning for image-based species identification. Meth. Ecol. Evol. 9, 2216–2225.
Ward, R.D., Zemlak, T.S., Innes, B.H., et al., 2005. DNA barcoding Australia's fish species. Phil. Trans. R. Soc. B 360, 1847–1857.
Ward, R.D., Hanner, R., Hebert, R.D.N., 2009. The campaign to DNA barcode all fishes. J. Fish Biol. 74, 329–356.
Watsa, M., Erkenswick, G.A., Pomerantz, A., Prost, S., 2020. Portable sequencing as a teaching tool in conservation and biodiversity research. PLoS Biol. 18, e3000667.
Webb, T.J., Mindel, B.L., 2015. Global patterns of extinction risk in marine and non-marine systems. Curr. Biol. 25, 506–511.
Webb, T.J., Vanden Berghe, E., O'Dor, R., 2010. biodiversity's big wet secret: the global distribution of marine biological records reveals chronic under-exploration of the deep pelagic ocean. PLoS One 5, e10223.
Wernberg, T., Krumhansl, K., Filbee-Dexter, K., Pedersen, M., 2019. Status and trends for the world's kelp forests. In: Sheppard, C. (Ed.), World Seas: An Environmental Evaluation: Ecological Issues and Environmental Impacts, second ed. vol. 3. Academic Press, London, UK, pp. 57–78.
Wesselmann, M., Geraldi, N.R., Duarte, C.M., Garcia-Orellana, J., Díaz-Rúa, R., et al., 2021. Seagrass (*Halophila stipulacea*) invasion enhances carbon sequestration in the Mediterranean Sea. Glob. Chang. Biol. 27, 2592–2607.

Wesselmann, M., Geraldi, N.R., Marbà, N., Hendriks, I.E., Díaz-Rúa, R., Duarte, C.M., 2022. eDNA reveals the associated metazoan diversity of Mediterranean seagrass sediments. Diversity 14, 549.
Wiens, J.J., 2015. Faster diversification on land than sea helps explain global biodiversity patterns among habitats and animal phyla. Ecol. Lett. 18, 1234–1241.
Wilkinson, C., Salvat, B., Eakin, C.M., et al., 2016a. Tropical and sub-tropical coral reefs. In: Iniss, L., Simcock, A., Ajawin, A.Y., et al. (Eds.), The First Global Integrated Marine Assessment. World Ocean Assessment I. United Nations Division for Ocean Affairs and Law of the Sea, New York, USA. 42 pp.
Wilkinson, M.D., Dumontier, M., Aalbersberg, I.J., et al., 2016b. The FAIR guiding principles for scientific data management and stewardship. Sci. Data 3, 160018.
Will, K.W., Rubinoff, D., 2004. Myth of the molecule: DNA barcodes for species cannot replace morphology for identification and classification. Cladistics 20, 47–55.
Will, K.W., Mishler, B.D., Wheeler, Q.D., 2005. The perils of DNA barcoding and the need for integrative taxonomy. Syst. Biol. 54, 844–851.
Wilson, E.O., 1985. The biological diversity crisis. Bioscience 35, 700–706.
Wilson, E.O., 1988. The current state of biodiversity. In: Wilson, E.O., Peter, F.M. (Eds.), Biodiversity. National Academy Press, Washington DC, USA, pp. 3–20.
Wilson, S.K., Graham, N.A.J., Polunin, N.V.C., 2007a. Appraisal of visual assessments of habitat complexity and benthic composition on coral reefs. Mar. Biol. 151, 1069–1076.
Wilson, M.F.J., O'Connell, B., Brown, C., et al., 2007b. Multiscale terrain analysis of multibeam bathymetry data for habitat mapping on the continental slope. Mar. Geod. 30, 3–35.
Wilson, D.S., Leifer, I., Maillard, E., 2015. Megaplume bubble process visualization by 3D multibeam sonar mapping. Mar. Petrol. Geol. 68, 753–765.
Winston, J.E., 2018. Twenty-first century biological nomenclature—the enduring power of names. Integr. Comp. Biol. 58, 1122–1131.
Woodall, L.C., Andradi-Brown, D.A., Brierley, A.S., et al., 2018. A multidisciplinary approach for generating globally comparable data on mesophotic, deep-pelagic and bathyal biological communities. Oceanography 31, 76–89.
Woolley, S.N.C., Tittensor, D.P., Dunstan, P.K., et al., 2016. Deep-sea diversity patterns are shaped by energy availability. Nature 533, 393–396.
Worm, B., Tittensor, D.P., 2018. A Theory of Global Biodiversity. Monographs in Population Biology 60. Princeton University Press, Princeton, NJ.
Worm, B., Barbier, E.B., Beaumont, N., et al., 2006. Impacts of biodiversity loss on ocean ecosystem services. Science 314, 787–790.
WWF, 2015. Reviving the Ocean Economy: The Case for Action—2015. World Wildlife Fund International, Gland, Switzerland, Geneva, Gland, Switzerland. 60 pp https://c402277.ssl.cf1.rackcdn.com/publications/790/files/original/Reviving_Ocean_Economy_REPORT_low_res.pdf?1429717323.
WWF, 2020. In: Almond, R.E.A., Grooten, M., Petersen, T., WWF (Eds.), Living Planet Report 2020-Bending the curve of biodiversity loss. Gland, Switzerland. 159 pp.
Yamahara, K.M., Preston, C.M., Birch, J., Walz, K., Marin III, R., et al., 2019. In situ autonomous acquisition and preservation of marine environmental DNA using an Autonomous Underwater Vehicle. Front. Mar. Sci. 6, 373.
Yoon, C.K., 2009. Naming Nature: The Clash Between Instinct and Science. W.W. Norton & Company Ltd., London, U.K. 341pp.
Zhang, Y., Kieft, B., Hobson, B.W., Raanan, B.-Y., Urmy, S.S., et al., 2019. Persistent sampling of vertically migrating biological layers by an autonomous underwater vehicle within the beam of a seabed-mounted echosounder. IEEE J. Oceanic Eng. 46, 497–508.

CHAPTER THREE

Review study on governance and international law for coastal and marine ecosystems in response to climate change: Social science perspective

Kailei Zhu[a] **and Jiayu Bai**[b,*,†]
[a]School of International Affairs and public Administration, Ocean University of China, Qingdao, Shandong, China
[b]School of law, Nankai University, Tianjin, China
*Corresponding author: e-mail address: gracefulgl@hotmail.com

Contents

1. Introduction 118
2. Overview 119
   2.1 Visual overview 119
   2.2 Literal overview 125
3. Theoretical research on coastal and marine ecosystems in response to climate change 128
   3.1 Theoretics of governance based on international relations 128
   3.2 Theoretics of governance based on economics 129
   3.3 Theories of the application and change of the international legal regimes 130
4. Methodological studies on coastal and marine ecosystems in response to climate change 131
   4.1 Management tools for coastal and marine ecosystems in response to climate change 131
   4.2 Policies for coastal and marine ecosystems in response to climate change 132
   4.3 International law on coastal and marine ecosystems in response to climate change 133
5. Empirical studies on coastal and marine ecosystems in response to climate change 133
   5.1 Empirical studies from a regional perspective 133
   5.2 Empirical Studies from a National Perspective 135
   5.3 Empirical studies from an institutional perspective 136
6. Discussion 137

† These authors contributed equally to this work and should be considered as co-first authors.

Advances in Marine Biology, Volume 93
ISSN 0065-2881
https://doi.org/10.1016/bs.amb.2022.09.003

6.1 Theoretical advancements in the response of coastal and marine ecosystems to climate change 137
6.2 Methodological advancements in the response of coastal and marine ecosystems to climate change 138
6.3 Empirical study trends of coastal and marine ecosystems responses to climate change 139
7. Conclusion 140
Acknowledgement 140
References 140

Abstract

As a common concern of humankind, the governance of coastal and marine ecosystems is increasingly coming to the fore of the international community as part of the joint response to climate change. Since the signing of the *Declaration of the United Nations Conference on the Human Environment* and the *United Nations Convention on the Law of the Sea* several decades ago, the international community has been exploring how international law can be improved in this respect. At present, the governance and international law of coastal and marine ecosystems in response to climate change are studied from theoretical and methodological perspectives. Extensive empirical studies help pinpoint specific issues related to each topic and provide valuable empirical references for both developed and developing countries. Based on social science publications, the authors use technical means to visualize research related to this topic, and conduct comprehensive reviews of these papers. They reveal that research based on these topics started late and is characterized by fragmentation. The research potential related to mentioned topic has yet to be explored extensively.

1. Introduction

At the 2022 United Nations Conference on the Law of the Sea, the Secretary-General of the United Nations pointed out that "the ocean connects all people...we are facing an 'ocean emergency,' where we must reverse the tide of rising sea levels and seawater acidification caused by ocean climate change (UN, 2022)." The discussion on the governance of joint response to climate change has long attracted considerable attention from the international community and in fact, this issue has been a concern of international law since the 1980s. According to data published on the website of the *United Nations Convention on the Law of the Sea* (UNCLOS), as early as 1989, the Secretary-General's report on the protection and preservation of the marine environment acknowledged the issue of a joint response to climate change. Both global governance and international law of the international community focus on the response of coastal and marine ecosystems to climate change. According to WWF research, if the ocean were a country, it would be the seventh largest economy in the world (WWF, 2015). Therefore, in the context of climate change, coast and ocean are undoubtedly of great significance to human survival and development. As one of the most pressing issues concerning the shared future for mankind, publications on governance and international law are increasing. From a perspective of governance and international law, the authors review the current state of social science research in relation to the response of coastal and marine ecosystems to climate change, and on this basis, propose research prospects.

2. Overview

2.1 Visual overview

As one of the most comprehensive bibliographic databases available, Web of Science (WOS) provides users with a convenient way to access publications online, including Science Citation Index Expanded (SCI-expanded), Social Science Citation Index (SSCI), and Art & Humanities Citation Index (A & HCI) (Wei et al., 2015). When searching WOS for publications related to this topic using a specific query formula, the authors find 432 results from these citation databases, including 341 publications containing papers and book chapters. Taking WOS's document categories as the standard, the authors analyze the first 25 categories of 341 selected publications, and find there is an extensive range of fields related to this topic. In addition to research focusing on environmental sciences and oceanography, it also involves humanities and social sciences such as international relations, economics, law, public administration, and sociology (Fig. 1).

It helps to the bibliometric relationship of a theme. Despite a large quantity of publications on this topic, few is actually known about its overall knowledge landscape. Domain visualization has attracted extensive attention due to various visualization tools, which perform domain analysis in publications (Boyack, et al., 2002). CiteSpace (6.1 R2) is an application developed by Chen Chaomei (Chen, 2004), and provides solutions for bibliometric analysis, data mining algorithms, and visualization. It can be used to identify knowledge bases, milestones, hotspots, emerging trends, and key points in a quantity of publications of the same kind. Compared with other visualization tools, this application is instrumental in improving the clarity and interpretability of visualization, helping users reduce cognitive burden, and finding important trends and key points in the total knowledge landscape (Chen, 2006), which contribute to its wide use in

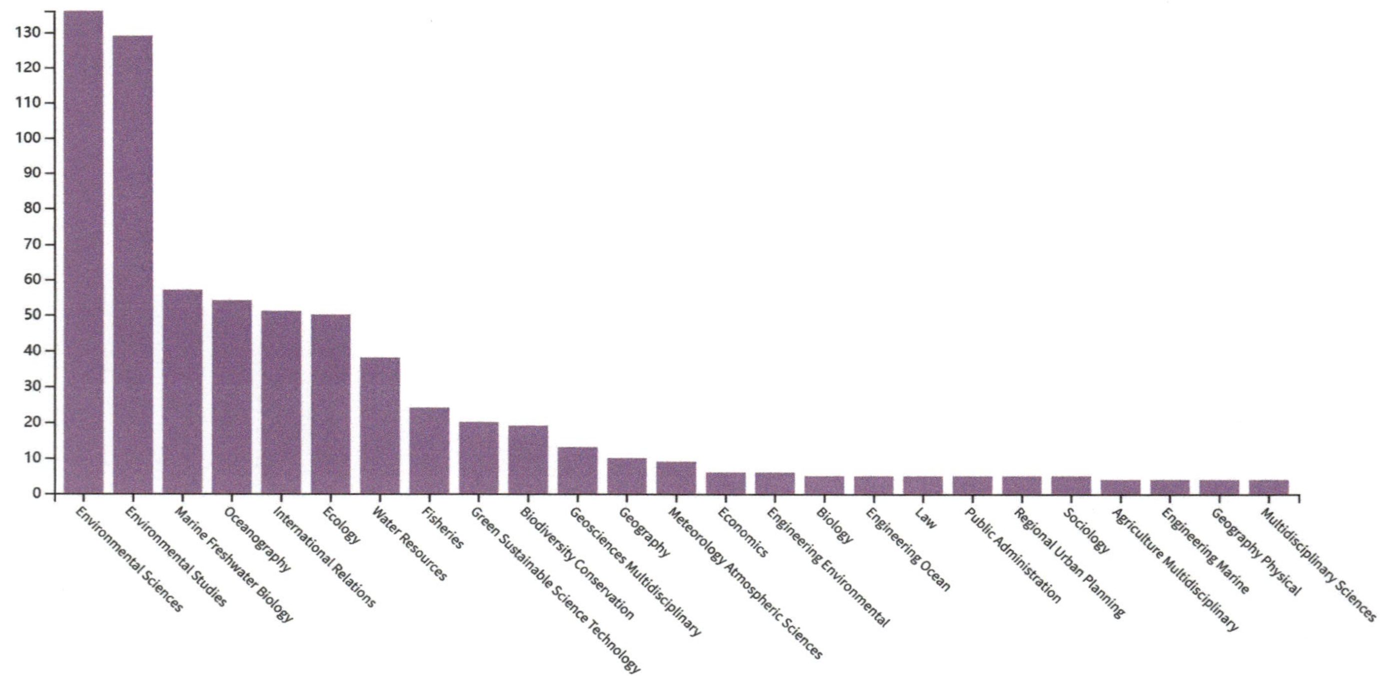

Fig. 1 Analyzed results (341 publications selected from SCI-Expanded, SSCI, and A&HCI within the Web of Science Core Collection).

various disciplines (Fang et al., 2018). In order to provide a systematic and objective overview of research on this topic, the authors select 71 potential publications that have been labelled as international relations, economics, law, public administration, and sociology according to the WOS disciplinary classification since July 2022 from the above 341 publications, all of which are marked as an article by WOS, including one article & proceedings paper and one article & early access, but no book chapters. These papers are visualized with CiteSpace software.

Using the default value, Fig. 2 shows the academic collaboration between authors with a threshold value of 1 for Node Labels. The network of author co-authorship, consisting of 215 nodes and 517 links, contributes to the joint response of coastal and marine ecosystems to climate change from governance and international law perspectives. The size of the node represents the quantity of theses published by the authors, whereas the distance between nodes and the thickness of links indicate the degree of collaboration between authors. As can be seen from Fig. 2, the network has a large quantity of participants and extensive collaboration, demonstrating the interdisciplinary nature of the topic. However, the whole network contains some isolated small subnetworks and single individuals, which means that authors in the field of social sciences tend to collaborate in small groups. Several authors are common contributors to this topic. In summary,

Fig. 2 Visualization of author collaboration network.

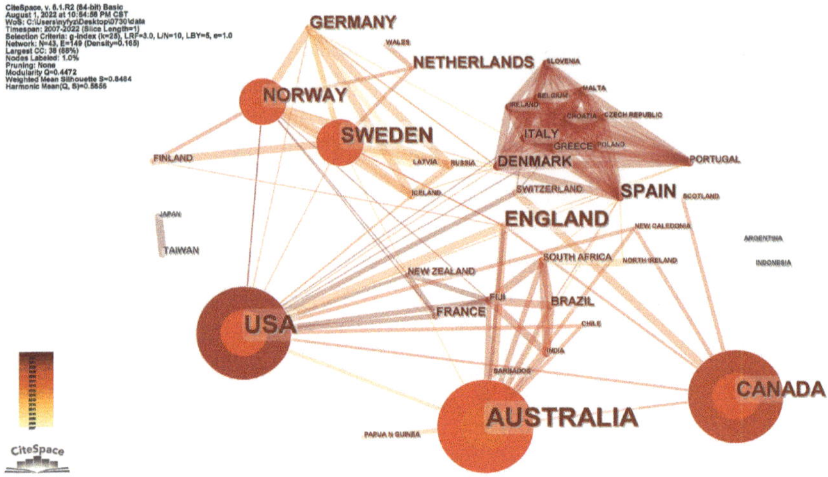

Fig. 3 Visualization of national collaboration network.

although social science research groups are typically smaller than science and engineering research groups, authors researching on this topic tend to collaborate in the form of groups.

These authors are distributed in different countries and regions. Fig. 3 shows the academic collaboration between authors with a threshold value of 1 for Node Labels, again using the default value of the software. The nodes represent the countries and regions, and their sizes represent the quantity of theses from different countries and regions, whereas the distance between nodes and the thickness of links indicates the degree of collaboration between countries and regions, and the circles of nodes represent the key points connecting different parts of the network. The larger the ring area, the higher the betweenness centrality of the nodes. As can be seen from Fig. 3, which consists of 43 nodes and 149 links, most authors are from five developed countries, including the United States, Australia, Canada, Norway, and Sweden, and the United States has a higher frequency of contacts between the country and other institutions. In other words, these five countries play a key role in research collaborations on this topic in 43 countries and regions.

A journal co-citation analysis helps the understanding of the landscape of the academic field, in which academic journals are the main means of communication (Hu et al., 2011). Two journals would be co-cited when, at least, one paper of each journal is listed in a citing paper's reference list (McCain, 1991). The authors draw Fig. 4 through CiteSpace, where nodes

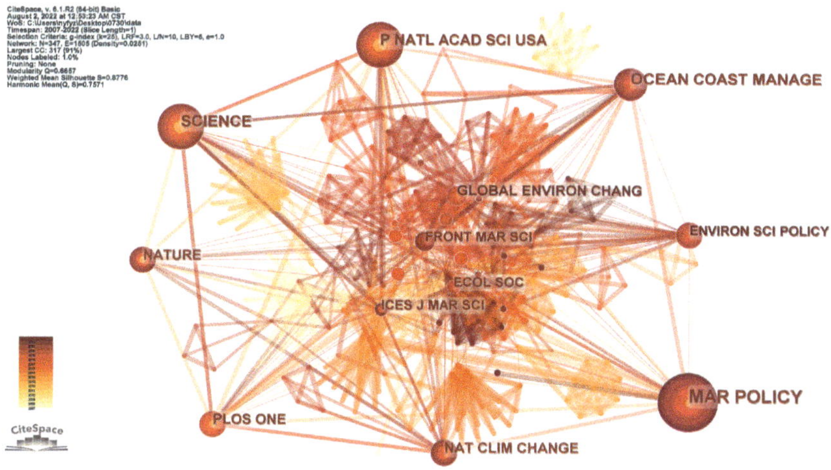

Fig. 4 Visualization of journal co-citation network.

represent journals, and links represent co-citation relationships between these journals. In general, the larger the node, the more important the journal. Fig. 4 shows that the distribution of journals is dominated by journals such as *Marine Policy* and *Science*, and the threshold value for node labels in this figure is 15, including 347 nodes and 1505 links. Both cited journals have impact factors greater than 3, and the journals are generally high-quality and reliable.

Information about the core content of the papers could be reflected by keywords. Over time, a knowledge map of keyword co-occurrences reveals popular topics, while burst keywords (keywords frequently cited over a period of time) indicate frontier topics (Yu et al., 2017). By running CiteSpace, the authors draw Fig. 5 and set the threshold value for the keyword to 20, where each node represents a keyword whose size is proportional to the co-occurrence frequency. Table 1 lists the top 20 burst keywords with the most robust citation in the fields related to the topic, with a γ value of 0.3, and among the 23 results presented thereafter, the first 20 are selected. The analysis shows that climate change (including mitigation and adaptation) and governance/management, as being keywords most closely related to the topic and receiving most attention. However, keywords related to coastal and marine ecosystems and international law are not precise, but are further extended to terms such as ecosystem services (ES), ecosystem-based management (EBM), fisheries, and frameworks, indicating that further systematic research on coastal and marine ecosystems is of great

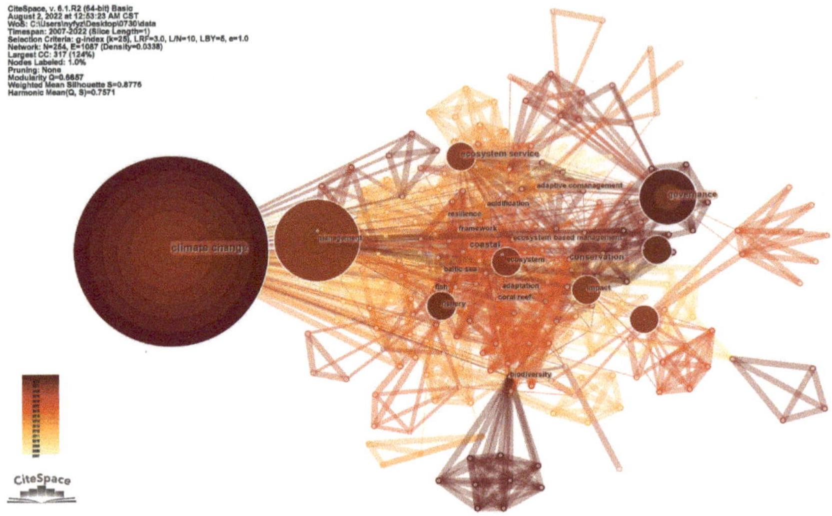

Fig. 5 Visualization of keyword co-occurrence network.

Table 1 Top 20 keywords with the strongest citation burst.

Cited Journals	Year	Strength	Begin	End	2007 - 2022
resilience	2007	1.35	**2007**	2016	
fish	2007	1.27	**2010**	2013	
great barrier reef	2007	0.96	**2010**	2013	
fishery	2007	1.84	**2012**	2014	
biodiversity	2007	1.59	**2012**	2013	
acidification	2007	1.53	**2012**	2014	
ecological system	2007	1.09	**2012**	2013	
climate change adaptation	2007	1.07	**2013**	2016	
ecosystem	2007	1.38	**2016**	2018	
ocean	2007	2.09	**2017**	2018	
impact	2007	1.13	**2017**	2019	
marine protected area	2007	1.03	**2017**	2018	
bluefin tuna	2007	0.95	**2018**	2019	
coastal	2007	1.46	**2019**	2020	
adaptation	2007	1.19	**2019**	2022	
management	2007	1.12	**2019**	2020	
ocean governance	2007	1.65	**2020**	2022	
ecosystem service	2007	1.64	**2020**	2022	
fisheries management	2007	1.47	**2020**	2022	
framework	2007	1.06	**2020**	2022	

value to joint response to climate change. This helps bridge existing marginal studies on this topic. Based on the analysis of the burst keywords in recent years, the authors hold that international law means studies can appropriately be termed the "frameworks", while "marine governance" covers the governance of coastal and marine ecosystems in the context of joint response to climate change. Therefore, it is reasonable to believe that the topic of this paper is in good agreement with the research hotspots in recent years.

2.2 Literal overview

These 71 papers are broadly grouped into theoretical, methodological, and empirical perspectives. Theoretical perspectives typically involve the study of general social science disciplines, including economics, international relations, and international law; methodological perspectives are divided into policy, management, and legal domains; empirical perspectives include country analysis, regional analysis, and institutional analysis. It is found that empirical analysis account for most of the papers (about 48% of the total number of papers), which shows a pragmatic attitude towards solving specific issues. The papers in theoretical perspectives and in methodological perspectives account for about 20% and about 24% of the total quantity of papers, respectively.

Papers related to theoretical studies generally involve governance based on international relations and governance based on economics and international law. First, the field of governance based on international relations focuses on the institutional change in the governance system (Österblom, et al., 2010) and the assessment of the adaptability of marine resource governance to environmental change (Dutra, et al., 2019) to avoid reaching critical points (Serrao-Neumann, et al., 2016) and advocate collective action (Berkowitz, et al., 2020), etc. Second, the field of governance based on economics focuses on the principles of sustainable governance of the coastal zone (Duxbury and Dickinson, 2007). Furthermore, in these papers, based on knowledge of welfare economics and blue financing, thorny issues in coastal and marine governance are studied (Groeneveld, 2020), and a conceptual framework for ocean protection and development is put forward (Shiiba, et al., 2022). Third, the field of international law focuses on issues such as standards for effective transboundary biodiversity conservation (Lim, 2016), the incorporation of climate change impacts into environmental impact assessment of marine areas beyond national jurisdiction (Warner, 2018), and marine zoning (Craig, 2012).

Papers related to methodological studies highlight many viewpoints on management, policy, and law. First, management-related studies account for the majority of these papers, and are further divided into three fields: coastal and marine ecosystems management and marine fisheries management. For the first of these, in the coastal-related studies, the challenges and opportunities faced by ecosystem management around the issue of "How Beach Managers Can Lead Transformative Change" (Corrêa et al., 2020) are assessed. In addition, the construction of the observation system framework in the field is also discussed (Malone, et al., 2014). In the marine field, the focus is on the sustainable management process of the marine environment (Scharin, et al., 2016), and the analysis of the necessity of marine iron utilization in geoengineering (Güssow, et al., 2010). At the same time, the field also calls attention to the threat of marine plastic pollution (Villarrubia-Gómez, et al., 2018), etc. For studies related to marine fisheries management, it concerns the effective management of the resources of regional fisheries management organizations (RFMOs) during climate change (Pentz, et al., 2018). It is shown that promoting adaptive marine aquaculture (Nursey-Bray, et al., 2018) can help better address the challenges and opportunities of fisheries (Sumaila, et al., 2016). Second, in terms of the policy-related studies, scientific decision-making (De Donà and Linke, 2022) and the construction of climate change regime (Guggisberg, 2019), among others, have played a crucial role in the ocean's response to climate change. Thirdly, for the studies related to international law, mainstream conventions such as the *United Nations Framework Convention on Climate Change* (UNFCCC) (Harrould-Kolieb, 2019) and the *Paris Agreement* (Dobush et al., 2022) have established the basic international legal framework.

As mentioned above, papers related to empirical studies make up the majority of the selected papers, covering regional, national, and institutional fields. Regionally relevant studies are the most abundant papers in empirical studies, and are further divided into regional marine fisheries, regional marine ecosystem, regional policy, and international law.

Firstly, regional marine fisheries-related studies focus on the analysis of artisanal fisheries management (Corral and de Lara, 2017) and small-scale fisheries (Hanich, et al., 2018) of the archipelago in the context of joint response to climate change, and it is believed that these are operable in the context of joint response to climate change. Regional marine fisheries-related studies also focus on climate change impacts, adaptation options, and outcomes in specific countries or regions, such as Australia

(Evans, et al., 2013; Chang, et al., 2013), and creatively put forward the idea of consensus management of the polar high seas (Nilsson, et al., 2016), etc.

Secondly, with regard to regional marine ecosystem-related studies, by summarizing the impacts of climate change in the United Kingdom, Canada, and other countries on the Canadian Pacific marine ecosystem (Talloni-Álvarez, et al., 2019), it is pointed out that there are currently perceived threats and constraints to environmental management related to marine biodiversity (Forster, et al., 2011). Navigating nature-based coastal adaptation through barriers is believed to be one of the viable approaches for coastal adaptation to climate change (Rahman, et al., 2021). This field focuses on the changes in the regulatory pattern of regional marine ecosystems in traditional regions such as the Baltic Sea region (Ringbom and Joas, 2018), and related issues in the high seas. It argues that global connectivity and cross-scale interactions create uncertainty for blue growth in Arctic fisheries (Niiranen et al., 2018), and calls for making of marine areas beyond national jurisdiction more resilient to climate change and other anthropogenic activities (Yadav and Gjerde, 2020), etc.

Thirdly, for regional policy and international law, the relevant papers conduct a risk assessment on marine and fisheries policy strategies in Europe based on a synthesis of current knowledge to reverse the downward trend (Bastardie and Brown, 2021). Some countries and regions have promulgated relevant policies and legislation on marine biodiversity (Saarman and Carr, 2013) and ocean acidification (Galdies, et al., 2020), and pay attention to the policy integration of fisheries (Khan, et al., 2018).

In terms of national-related studies, they involve not only marine fisheries (Ogier, et al., 2016) and marine ecosystems (Rogers, et al., 2016) in developed countries, but also marine fisheries (Van der Geest, 2017; Porobic et al., 2018; Belhabib, et al., 2019; Truchet, et al., 2022) and coastal ecosystems in developing countries (Aziz, et al., 2016; Mursyid, et al., 2021; Văidianu, et al., 2020).

Studies on regional institutions are divided into fisheries institutions, community institutions, and natural park institutions. In the related studies of fisheries institutions, the response of regional fisheries institutions to climate change is assessed (Sumby, et al., 2021), and it is believed that CCAMLR could adapt to the challenges brought about by climate change (Wendebourg, 2020). Meanwhile, community institutions-related studies call for building resilience to natural disasters through coastal governance (Dunning, 2020), and point to the need to involve coastal communities in the circumpolar north in the formulation and implementation of the approaches used to respond to climate change (Bonnett and Birchall, 2020). Whereas, studies related to natural park institutions involve the integration of ecology into tourism to build resilience (Jones, 2021), and focus on the balance of stakeholders in natural park management (Ricart and Rico-Amorós, 2022).

These papers are gradually building an increasingly mature landscape of research. It should be noted, however, that there are fewer papers from the perspective of international law than from the perspective of governance, and there appears to be a gap in the field of governance and international law integration. This provides a prerequisite for the innovation of subsequent research. The following is an overview of this knowledge landscape, and introduces the theoretical, methodological, and empirical-related studies divided from the papers.

3. Theoretical research on coastal and marine ecosystems in response to climate change

3.1 Theoretics of governance based on international relations

Regarding international relations, coastal and marine ecosystems are often regarded as a whole. Climate change is a major driver of changing the abundance and distribution of coastal and marine living resources, which negatively affects communities that depend on coastal and marine resources (Dutra, et al., 2019). An accurate and timely understanding of the status and trends of coastal and marine ES (benefits) will help address the challenges posed by increasingly intensive climate change to the coastal and marine areas (Teneva et al., 2022). By creating the Global Seafood Market Performance Index to assess the coastal and marine ES (benefits), it is found that the global driving factors such as seafood market, and climate change are collectively affecting the biodiversity of biological resources in coastal and marine ecosystems (Villasante, et al., 2012). In fact, the combined pressures of climate change and other negative drivers lead to the collapse of coastal and marine ecosystems, and these negative drivers are triggering a critical point that challenges the successful implementation of existing governance arrangements at the local level (Serrao-Neumann, et al., 2016). The joint development of science, policy, and practice is required to effectively reduce the cumulative impact of these negative drivers on coastal and marine ecosystems. Specifically, bottom-up governance institutional innovation from local to regional and EBM are crucial to solve this issue

(Österblom, et al., 2010). Furthermore, it is necessary to pay attention to the role of cross-sectoral or multi-stakeholder multilateral organizations in governance. The use of meta organizations (MO, i.e., organizations of organizations) to facilitate governance of marine complex issues including coastal and marine ecosystems in response to climate change is an innovative and beneficial theoretical model (Berkowitz, et al., 2020).

3.2 Theoretics of governance based on economics

The theories related to governance based on economics are further divided into coastal and marine-related studies. In terms of coastal-related studies, the ecosystem and infrastructure of the region are under increasing pressure against the backdrop of the global population increasing in coastal areas. Due to the neglect of the four major factors that enhance sustainability, nature, architecture, society, and human resources, many difficulties have arisen in the coastal areas, such as the growth of coastal population and the degradation of natural capital. Therefore, it is necessary to develop new strategies using the sustainable governance principles as a framework so that coastal communities can continue to live in these areas without further reducing their natural capital. Development of economic, human resources, and social capital, emphasis on natural capital, improved stakeholders' engagement, and integration of various actors will also facilitate successful application of the principles to fully prepare coastal areas for future coastal disasters (Duxbury and Dickinson, 2007).

In terms of marine related studies, ensuring adequate funding to sustain marine ecosystems and resources in the face of the adverse impacts of climate change and intense human activities is an ongoing challenge. Blue finance is a key tool in ensuring marine sustainability, and the use of legal and institutional mechanisms to increase investment in marine development projects holds great potential. Existing blue financing practice, however, suggests that the most important challenge is to establish links among all stakeholders to illustrate the current potential benefits of increased private investment and public contributions. Establishing regulatory-driven financing mechanisms and incorporating the core concept of blue finance into marine governance at the national and international levels are valuable in addressing such challenges. This approach promotes blue financing by maintaining a balance between economic growth and marine sustainability (Shiiba, et al., 2022). At the same time, definitional difficulties, uncertainty, and conflicting views and values remain as the thorny issue in coastal and marine governance. As an application of welfare economics, social cost-benefit analysis can be used to solve these issues by setting analytical assumptions (Groeneveld, 2020).

3.3 Theories of the application and change of the international legal regimes

In light of the negative impacts of climate change on coastal and marine areas, we call for the theoretics based on international law to quickly fill the gaps in respects of environmental protection, joint responses to climate change, and biodiversity conservation. For coastal and marine environmental protection, adverse effects of climate change have increased pressure on species, habitats, and ecosystems in all coastal and marine areas, and weakened the ecological services they provide. Determining the nature and extent of the impacts of climate change on coastal and marine biodiversity through environmental impact assessments and associated mitigation measures is a key step in reducing adverse impacts and halting biodiversity loss. Given the fact that the marine areas beyond national jurisdiction constitute the vast majority of the ocean, it is of great significance to address this challenge by incorporating the impact of climate change into the environmental impact assessment process through a new legally binding international instrument on the conservation and sustainable use of marine biodiversity in areas beyond national jurisdiction (Warner, 2018).

As for joint response to climate change, the increasingly severe climate change and its complex and synergistic impacts on coastal and marine ecosystems prove that existing governance systems have failed to protect coastal and marine ecosystems. Marine spatial planning is widely recognized as a means of improving the above through ecosystem-based overall planning. However, the concept was proposed earlier, when the joint response to climate change was not significant and cannot adequately explain the dynamic changes in marine ecosystems caused by climate change. Anticipatory zoning and encouraging private participation allow flexibility in marine spatial planning to adapt to changes that climate change will bring to coastal and marine areas (Craig, 2012).

In terms of biodiversity, as the world continues to suffer from the impacts of climate change, the proper management of intact ecosystems beyond jurisdictional boundaries is increasingly important. However, ecosystems are often artificially divided by administrative divisions and are therefore restricted by uncoordinated management across countries. In conjunction with biodiversity conservation laws, creating an indicator to assess the potential effectiveness of transboundary conservation systems in adapting to climate change and protecting biodiversity has positive implications (Lim, 2016).

Theoretics based on international law need to not only fill the above-mentioned gaps, but also to confront the negative impacts of climate change on established stable theories. Climate change has objectively caused coastlines to retract. If boundaries in the natural environment are referred to as the statutory boundaries, maritime boundaries may be displaced (Árnadóttir, 2021), and subsequently, court cases regarding maritime boundaries increase. Therefore, it is necessary to consider sustainable and equitable coastal and ocean governance in future trials of relevant cases at the time of judgement (Ch and Gupta, 2022).

4. Methodological studies on coastal and marine ecosystems in response to climate change

4.1 Management tools for coastal and marine ecosystems in response to climate change

Management tools for coastal and marine ecosystems in response to climate change are divided into related studies of coastal ecosystem management, marine ecosystem management, and marine fisheries management. For studies related to coastal ecosystem management, the sophisticated management of coastal ecosystems is based on coastal ecosystem data provided by the global marine observing system framework (Malone, et al., 2014). Based on observations, beaches are found to provide a range of ESs that are increasingly affected by climate change and other stress factors. EBM is a tool to respond to changing environments and ensure long-term ES. By integrating all coastal EBM into existing governance systems, local beach managers can facilitate evidence-based implementation of improving coasts (Corrêa, et al., 2020).

As for marine ecosystem management-related studies, it is necessary to promote sustainable management with an integrated approach to address the negative impacts of climate change (Scharin, et al., 2016) and the intertwined impacts of marine plastic pollution and climate change (Forster, et al., 2017) leading to continued deterioration in coral reef health under external pressures on marine ecosystems caused by climate change (Villarrubia-Gómez, et al., 2018). Compared with other emission reduction options, the potential of marine iron utilization cannot be ignored. At the same time, it is valuable to explore the positive implications of geoengineering for marine ecosystems in response to climate change, such as marine iron utilization.

In terms of marine fisheries management-related studies, ocean temperature and acidification caused by climate change have a negative impact on the health of fisheries populations, which is not conducive to the sustainable development of fisheries (Sumaila, et al., 2016). A survey of fishermen found that fish stocks had dwindled, and reforms were urgently needed. For example, licenses should be reduced, and regulation of recreational fishing should be strengthened (Nursey-Bray, et al., 2018). In the reform, RFMOs have played an important role in addressing fluctuations in fisheries resources caused by climate change. An assessment of RFMOs brought out that the policy framework of RFMOs is comprehensive and appears to have most of the elements required to achieve fisheries resources management objectives under climate change (Pentz, et al., 2018).

4.2 Policies for coastal and marine ecosystems in response to climate change

Marine climate policy plays an important role in mitigating and adapting to the negative impacts of climate change on marine ecosystems. The link between marine climate policy and marine biodiversity must be addressed (Dobush et al., 2022). When formulating forward-looking climate change policies, it is important to understand the contribution of the ocean to the global climate and to comprehensively review the services the ocean provides to mankind (Galland, et al., 2012). It calls for policymakers to make scientific decisions that not only bring science close to policy and politics, but also avoid being too close to avoid politicization. Specifically, the relationship between science and policy needs to be constructed along a continuum consisting of three levels: a clear-cut boundary, a boundary zone, and a fuzzy boundary. According to this, it is necessary for executive organizations to seize new learning opportunities and to balance credibility, legitimacy, and salience (De Donà and Linke, 2022). Furthermore, because developing countries rely heavily on the products provided by marine ES for food security and livelihoods, they are more vulnerable to the negative impacts of climate change on marine ecosystems. Consequently, it is necessary to particularly support the adaptation of developing states under the climate change regime (Guggisberg, 2019).

4.3 International law on coastal and marine ecosystems in response to climate change

Most studies related to international law focus on the legal analysis of coastal and marine fisheries. Pursuing potential fishing opportunities presents considerable management challenges, and emerging fisheries too are highly vulnerable to the risk of overfishing. The combination of an effective regulatory vacuum and the prospect of high short-term profits has contributed to the boom-and-bust cycle. International law imposes substantial requirements for coastal and marine fisheries to minimize the negative impacts of climate change and increase the overall resilience of coastal and marine fisheries (Craig, 2019). Therefore, the contribution of the UNFCCC and the *Paris Agreement* to the response to climate change of coastal and marine fisheries cannot be ignored (Harrould-Kolieb, 2019). It is necessary to further explore the scope of legal obligations involved in coastal and marine fisheries and to collate developing international practices (Caddell, 2018). For policymakers, it is imperative to support the legal and policy tools, processes, and institutions required for ecosystem-based ocean management (Macpherson, et al., 2021).

5. Empirical studies on coastal and marine ecosystems in response to climate change

5.1 Empirical studies from a regional perspective

Empirical-related studies from a regional perspective are divided into the studies of regional fisheries, regional ecosystems, regional policies, and international law. As far as regional fisheries-related studies are concerned, it is further divided into the studies of coastal and marine. Firstly, as the coastal zone has the closest relationship with human production and life, it is one of the most suitable areas for empirical research, especially those areas where islands and rocks are widely present. It has been proven that the socioeconomic development of fishing communities with vast islands and rocks, to a large extent, depends on local coastal and marine resources (Corral and de Lara, 2017). For them, environmental changes have always been transient and reversible. However, climate change is an important driving force for changes in coastal social ecosystems (Evans, et al., 2013). At present, climate change has impacted and is expected to continue to non-linearly affect national territories where islands and rocks are widely present, including rising temperatures, rising sea levels, bleaching coral reefs and marine acidification (Hanich, et al., 2018). Fishing activities around islands and rocks are plagued by extreme weather events (Chang, et al., 2013). Also, it is undeniable that small-scale fisheries are critical to global livelihoods, food security, and income. Therefore, various forms of spatial management are widely adopted in small-scale fisheries as potential solutions to these complex issues (Le Cornu et al., 2018). In addition, alternative livelihood strategies can be employed to generate income while protecting fish stocks and their habitats, such as the development of recursive catch and release fishing for game fish (Wood, et al., 2013). Secondly, the exploration of marine fisheries has never stopped. In face of the fact that climate change is affecting marine regions, including the Southern Ocean, it is necessary for governance institutions to accelerate the pace of decision-making in relevant regional ecosystems in response to climate change (Nilsson, et al., 2016).

The studies on regional ecosystems are divided into regional ecosystems within national jurisdiction and that beyond national jurisdiction, where the former mainly refers to coastal areas. Based on observations and perceptions, the impacts of climate change on coastal ecosystems in different regions are found to be different. In the British Overseas Territories in the Caribbean, climate change is regarded as the greatest threat to the marine ecosystems in the future, including hurricane frequency, rise of sea surface temperatures, storm surges, changes in rainfall, and sea level rise (Forster, et al., 2011). In southern British Columbia, changes in fish stock distribution and fish abundance under climate change could have a significant impact on fish supply, and the extent of these impacts may vary with latitudinal gradient (Talloni-Álvarez, et al., 2019). Consequently, many international, regional, and national commitments have been made to restore coastal ecosystems in response to climate change (Yates, et al., 2013), involving insurance of adequate monitoring, participatory governance, and appropriate management plans (De Oliveira Soares and Lucas, 2018), such as nature-based coastal adaptation measures to help minimize the impacts of rising sea level in a sustainable manner to enhance the natural capacity of coastal ecosystems (Rahman, et al., 2021). At the same time, it is necessary to incorporate social factors into coastal governance (Carpenter, et al., 2018) and then build a land-ocean governance network (Pittman and Armitage, 2019). The implementation of the above-mentioned measures depends on the governance institutions with strong functions and close links between key regulators (Ringbom and Joas, 2018). As for studies related to regional ecosystems beyond national jurisdiction, climate change has created uncertainty about the blue growth of Arctic high seas fisheries (Niiranen et al., 2018) and limited the effectiveness of management tools for deep-sea areas in the North Atlantic (Johnson, et al., 2018). For good governance of marine fisheries resources (Kaiser, et al., 2015), it is necessary to employ dynamic tools for large-scale marine systems (Fidelman, et al., 2012), including making marine areas beyond national jurisdiction more resilient to climate change and other anthropogenic activities (Yadav and Gjerde, 2020).

Regional policy-related studies mainly focus on coastal fisheries policy. Around the world, ecosystem approaches-based fisheries management policies are being developed, and enhanced fisheries regulation could partially mitigate the impacts of climate change on fisheries level in coastal ecosystems and reduce the risk of biodiversity loss (Bastardie and Brown, 2021). However, it is a challenge to link sectoral tools to local adaptation policies, and more efforts are needed to identify opportunities for complementary adaptation strategies and policy integration to foster multiple interests (Khan et al., 2018).

The studies related to international law focus primarily on legislative issues for protection and management of marine ecosystems. The negative impacts of climate change, such as the effects of ocean acidification on marine ecosystems, require strong legislation to support governance (Galdies, et al., 2020). Therefore, it is necessary to adopt legislation to determine the management procedures of marine protected areas and temporary conservation measures for fisheries resources in certain high seas areas (Saarman and Carr, 2013). In addition, in the future, efforts can also be put into promoting the implementation of legislation through international cooperation (Hoel, 2020).

5.2 Empirical Studies from a National Perspective

Empirical studies from a national perspective include the studies on both developed and developing countries. For developed countries, the focus is on Australia's coastal fisheries and marine ecosystems. In the case of the former, good governance of fisheries is essential for building adaptability and resilience of fishing. Australia has developed a series of fisheries management tools (i.e., EBM, adaptive management, etc.) and applied them in different fields (Ogier et al., 2016). As for the latter, marine coastal wetlands such as mangroves and coastal salt marshes provide abundant ES, but climate change has exacerbated the loss of these. Therefore, explicit recognition of ES in climate legislation is urgently needed to protect these areas (Rogers et al., 2016).

Studies relevant to developing countries are vulnerable to increased attention due to their fragile ecosystems and lack of capacity and funding. They focus on coastal areas in developing countries and are divided into the studies of coastal fisheries and coastal ecosystems. With regard to studies related to coastal fisheries, a large quantity of small-scale fisheries in developing countries lack ecological knowledge (Truchet et al., 2022), which makes it difficult to address the threats posed by climate change. Therefore, there is an urgent need to redesign fisheries governance mechanisms (Van der Geest, 2017). However, implementing an ecosystem-based approach to fisheries management faces challenges in financial, capacity, and other aspects. The fisheries sector is plagued by policy mistakes and illegal activities (Belhabib et al., 2019). Therefore, to implement ecosystem-based fisheries management, there must first be reforms. It is necessary to establish management and scientific committees around ecosystems rather than single species, involving a wider range of stakeholders (Porobic et al., 2018). As for the coastal ecosystem-related studies, human intervention on the coast over the past half century has exacerbated climate change, altered the natural trend of coastal evolution, and had a significant impact on the coastal ecosystem. Considering the carbon sink capacity of mangroves for coastal ecosystems in response to climate change (Aziz, et al., 2016), mangrove conservation activities can be considered as the starting point (Mursyid et al., 2021). However, policies and international laws related to coastal ecosystem governance are not well enforced in some developing countries. The legislations are sectoral, with overlapping authorities and decentralized responsibilities. There is a gap between the contents of policies and reality (Văidianu et al., 2020). Therefore, it is necessary to pay attention to and support developing countries in the long term to reverse this situation.

5.3 Empirical studies from an institutional perspective

Empirical studies from the institutional perspective are further divided into the studies of fisheries institutions, community institutions, and natural park institutions. As for the fisheries institutions-related studies, coastal and marine changes caused by climate change will continue to and irreversibly alter existing fisheries conditions, posing challenges for fisheries management (Sumby et al., 2021). At present, the activities of administrative organizations for fisheries are based on international laws. Taking the Southern Ocean as an example, from an international perspective, most of the waters around Antarctica fall under UNCLOS, while the Antarctica fisheries are regulated from a regional perspective by the *Convention for the Conservation of Antarctic Marine Living Resources* (Wendebourg, 2020).

As mentioned above, in community institutions-related studies, coastal ecosystems responses to hurricanes, a representative sign of climate change, has drawn widespread attention. Accordingly, it is necessary to improve the resilience of coastal community ecosystems to natural disasters. One study found that participatory multi-stakeholder processes improved stability through disaster simulation exercises and participatory planning processes (Dunning, 2020). Considering the unique and fragile ecosystems in some regions such as the Arctic, to enhance resilience, it is necessary for some coastal community ecosystems to adopt a diverse portfolio of adaptations that incorporate more sustainable non-structural and ecosystem-based adaptation tools (Bonnett and Birchall, 2020).

As for the natural park institutions-related studies, the core issue is the conservation and rational utilization of coastal ecosystems in response to climate change. On the one hand, taking into account the original intention of establishing natural parks and other issues, conservation contributes to the full restoration of coastal ecosystem (Ricart and Rico-Amorós, 2022). On the other hand, considering the differences in the capacity of stakeholders in representativeness, relevance, and cooperation, it is also reasonable to adopt rational utilization-based incentive measures to promote the achievement of conservation objectives (Jones, 2021).

6. Discussion

6.1 Theoretical advancements in the response of coastal and marine ecosystems to climate change

In recent years, studies on governance theory based on international relations have developed vigorously. These studies include the evaluation of governance theory based on international relations and the innovation of governance theory. Although current studies have enhanced the richness of the fields covered by the content due to multiple perspectives, it shows fragmentation. This feature reflects that climate change and ecosystem factors are complex. For the future, it is worth looking forward to further exploration of a systematic governance theory. It is worth nothing that this conclusion may be limited, possibly due to the incomplete types of publications collected in this review.

Compared with the former, the theoretical studies on governance based on economics started earlier, but they are still insufficient. In fact, economics appears to be on the verge of addressing how coastal and marine ecosystems are currently responding to climate change. The reason for this phenomenon is that the core issues, that is, collaboration based on good international relations and good compliance with international law, have not been properly addressed. However, the studies on governance based on economics have potential, as any cooperation and compliance issues need to be addressed by making good use of the operation of laws of economic society.

Meanwhile, studies related to international law emerged later than governance studies based on international relations, which is also consistent with the fact that international law is a secondary focus of one of the tools of governance, even though international law itself has its own independent value. In recent years, the quantity of papers related to international law ranks second. However, considering that there are many international laws in coastal and marine environmental protection, joint responses to climate change, and biodiversity conservation, there is a lot of potential research space in studies of international law. In other words, the current quantity of papers in this field does not conform to the abundant rules and regulations of international law in this field, and there is potential for further exploration. Furthermore, the rule of law is vital to achieving good governance. Therefore, it is necessary to strengthen the study of international law for in-depth studies of international laws of various tools to address climate change.

6.2 Methodological advancements in the response of coastal and marine ecosystems to climate change

As the fundamental link in the methodology, studies on management methods occupy an important position. As retrieved, studies show strong development trends and broad scope. However, it is necessary to point out that these studies appear to be in their infancy. Evidence in support of this conclusion is that the topics reflected in the titles of these papers are primarily exploratory assessments or cutting-edge concepts. Unlike laws and policies, general management approaches ultimately concern solving specific issues. The current exploratory research lays a foundation for further development in this field to solve specific issues.

Papers related to policy tools construct a picture of the unfolding knowledge tree, and policies have advantages over laws in terms of flexibility and quantity. The policy tool is abstract, which is based on specific policies and guides the entire process of formulation, implementation, and supervision of policies. Studies of this kind are rare, and further research into policy tools by researchers could help improve the current situation. However, the abundance of papers related to policy tools is often related to the abundance of papers related to empirical policy analysis, sometimes based on the summaries of empirical research. Considering that papers related to empirical studies are still in the initial stage, it is expected that a substantial growth of the papers related to research on policy tools will take time.

Papers related to international law studies are more updated than those related to policy tools. This conclusion is drawn through the analysis of the time of papers. Papers related to tools of international law are all from the previous 4 years, and papers related to policy tools date back 10 years before. Climate change does pose many new challenges to international law related to this topic. Combined with the international context of the first part of this review, it is not difficult to understand that any papers related to the tools of international law are new as they are cutting-edge.

6.3 Empirical study trends of coastal and marine ecosystems responses to climate change

Regional studies are diverse with a wide range of contents. There are more regional studies than national studies. The key to the response of coastal and marine ecosystems to climate change lies in the ecosystems themselves. Therefore, the ecosystem-based approach plays a vital role in addressing related issues. However, it is well known that the boundaries of ecosystems and the boundaries of administrative divisions are often misplaced. As a solution, the ecosystem-based approach frequently presented in this review fully respects the ecosystem's boundaries. The quantity of regionally relevant empirical research proves that the empirical studies fully respect and implement the ecosystem-based approach.

National-related studies analyze developed countries less than developing countries. To a certain extent, it reflects the concern of the international community for developing countries. However, as developed countries grow the governance capacity to further explore their experiences and lessons related to this topic, it is preferable for developing countries to further explore developed countries' experiences and lessons related to this topic. As mentioned above, although regional ecological orientation is fundamental, national analysis is essential.

Institutional analysis is not mainstream, but it cannot be ignored that institutions play an important role in addressing issues related to this topic. Institutions help concentrate the strengths of countries in a particular region and are effective in solving regional issues related to this topic. However, there is a lack of equivalence between facts and research. For example, questions about fisheries account for a large part of this review, but there is little research on solutions for fisheries-related institutions. Therefore, a comprehensive analysis of all institutions relevant to the topic is required to reduce the extent of this asymmetry.

7. Conclusion

The Earth is a delicate and complex system, and has remained relatively stable for the past 10,000 years. In recent decades, however, human activities have begun to break this stable equilibrium and become the dominant force shaping the Earth's surface, so since 2020, artificial mass has surpassed the natural biomass for the first time (Elhacham, et al., 2020). However, human beings still know very little about the consequences of this massive force on Earth, so six of the nine critical biophysical processes within the framework planetary boundary framework have been crossed over the past few decades (Nagarajan, 2022), indicating that part of the damage caused by human beings to the planet is beyond the limits of most fields. Even at a high cost, recovery is not easy. In these astounding data, coastal and marine ecological issues under climate change, as a rapidly changing and non-linear issue, have challenged corresponding international legal governance. Papers based on the topic of this review started late, and there is ample room for research. These challenges can only be accomplished with the world's collective attention, and all countries of the world have a responsibility and obligation to work together to address climate change, protect coastal and marine ecosystems, and weave the blueprint of a community with a shared future for mankind.

Acknowledgement

This research is supported by the National Social Fund of China [Grant No.18VHQ001].

References

Árnadóttir, S., 2021. Fluctuating boundaries in a changing marine environment. Leiden J Int Law. 34, 471–487.

Aziz, A.A., Thomas, S., Dargusch, P., Phinn, S., 2016. Assessing the potential of REDD+ in a production mangrove forest in Malaysia using stakeholder analysis and ecosystem services mapping. Mar. Policy 74, 6–17.

Bastardie, F., Brown, E.J., 2021. Reverse the declining course: a risk assessment for marine and fisheries policy strategies in Europe from current knowledge synthesis. Mar. Policy 126, 104409. https://doi.org/10.1016/j.marpol.2021.104409.

Belhabib, D., Sumaila, U.R., Le Billon, P., 2019. The fisheries of Africa: exploitation, policy, and maritime security trends. Mar. Policy 101, 88–92.
Berkowitz, H., Crowder, L.B., Brooks, C.M., 2020. Organizational perspectives on sustainable ocean governance: a multi-stakeholder, meta-organization model of collective action. Mar. Policy 118, 104026.
Bonnett, N., Birchall, S.J., 2020. Coastal communities in the circumpolar north and the need for sustainable climate adaptation approaches. Mar. Policy 121, 104175. https://doi.org/10.1016/j.marpol.2020.104175.
Boyack, K.W., Wylie, B.N., Davidson, G.S., 2002. Domain visualization using VxInsight® for science and technology management. J Am Soc Inf Sci Tec. 53, 764–774.
Caddell, R., 2018. Precautionary management and the development of future fishing opportunities: the international regulation of new and exploratory fisheries. Int J Mar Coast Law. 33, 199–260.
Carpenter, A., Shellock, R., von Haartman, R., Fletcher, S., Glegg, G., 2018. Public perceptions of management priorities for the English Channel region. Mar. Policy 97, 294–304.
Ch, M.C.G., Gupta, J., 2022. Environmental and sociocultural claims within maritime boundary disputes. Mar. Policy 139, 105043. https://doi.org/10.1016/j.marpol.2022.105043.
Chang, Y., Lee, M.A., Lee, K.T., Shao, K.T., 2013. Adaptation of fisheries and mariculture management to extreme oceanic environmental changes and climate variability in Taiwan. Mar. Policy 38, 476–482.
Chen, C., 2004. Searching for intellectual turning points: progressive knowledge domain visualization. Proc. Natl. Acad. Sci. U.S.A. 101, 5303–5310.
Chen, C., 2006. CiteSpace II: detecting and visualizing emerging trends and transient patterns in scientific literature. J Am Soc Inf Sci Tec. 57, 359–377.
Corral, S., de Lara, D.R.M., 2017. Participatory artisanal fisheries management in islands: application to the Canary Islands (Spain). Mar. Policy 81, 45–52. https://doi.org/10.1016/j.marpol.2017.03.011.
Corrêa, M.R., Xavier, L.Y., Holzkamper, E., Andrade, M.M.D., Turra, A., Glaser, M., 2020. Shifting shores and shoring shifts-how can beach managers lead transformative change?: a study on challenges and opportunities for ecosystem-based management. Hum. Ecol. Rev. 26, 59–84.
Craig, R.K., 2012. Ocean governance for the 21st century: making marine zoning climate change adaptable. Harv. Envtl. L. Rev. 36, 305–350.
Craig, R.K., 2019. Fostering adaptive marine aquaculture through procedural innovation in marine spatial planning. Mar. Policy 110, 103555.
De Donà, M., Linke, S., 2022. 'Close but not too close'–experiences of science-policy bridging in three international advisory organizations. Crit. Policy St. https://doi.org/10.1080/19460171.2022.2028173.
De Oliveira Soares, M., Lucas, C.C., 2018. Towards large and remote protected areas in the South Atlantic Ocean: St. Peter and St. Paul's archipelago and the Vitória-Trindade seamount chain. Mar. Policy 93, 101–103. https://doi.org/10.1016/j.marpol.2018.04.004.
Dobush, B.J., Gallo, N.D., Guerra, M., Guilloux, B., Holland, E., Seabrook, S., Levin, L.A., 2022. A new way forward for ocean-climate policy as reflected in the UNFCCC Ocean and climate change dialogue submissions. Clim. Policy 22, 254–271. https://doi.org/10.1080/14693062.2021.1990004.
Dunning, K.H., 2020. Building resilience to natural hazards through coastal governance: a case study of hurricane Harvey recovery in Gulf of Mexico communities. Ecol. Econ. 176, 106759.
Dutra, L.X., Sporne, I., Haward, M., Aswani, S., Cochrane, K.L., Frusher, S., Gasalla, M.A., Gianesella, S.M., Grant, T., Hobday, A.J., Jennings, S., 2019. Governance mapping: a

framework for assessing the adaptive capacity of marine resource governance to environmental change. Mar. Policy 106, 103392.

Duxbury, J., Dickinson, S., 2007. Principles for sustainable governance of the coastal zone: in the context of coastal disasters. Ecol. Econ. 63, 319–330. https://doi.org/10.1016/j.ecolecon.2007.01.016.

Elhacham, E., Ben-Uri, L., Grozovski, J., Bar-On, Y.M., Milo, R., 2020. Global human-made mass exceeds all living biomass. Nature 588, 442–444.

Evans, L.S., Hicks, C.C., Fidelman, P., Tobin, R.C., Perry, A.L., 2013. Future scenarios as a research tool: investigating climate change impacts, adaptation options and outcomes for the great barrier reef, Australia. Hum. Ecol. 41, 841–857.

Fang, Y., Yin, J., Wu, B., 2018. Climate change and tourism: a scientometric analysis using CiteSpace. J. Sustain. Tour. 26, 108–126. https://doi.org/10.1080/09669582.2017.1329310.

Fidelman, P., Evans, L., Fabinyi, M., Foale, S., Cinner, J., Rosen, F., 2012. Governing large-scale marine commons: contextual challenges in the coral triangle. Mar. Policy 36, 42–53.

Forster, J., Turner, R.A., Fitzsimmons, C., Peterson, A.M., Mahon, R., Stead, S.M., 2017. Evidence of a common understanding of proximate and distal drivers of reef health. Mar. Policy 84, 263–272.

Forster, J.O.H.A.N.N.A., Lake, I.R., Watkinson, A.R., Gill, J.A., 2011. Marine biodiversity in the Caribbean UK overseas territories: perceived threats and constraints to environmental management. Mar. Policy 35, 647–657.

Galdies, C., Bellerby, R., Canu, D., Chen, W., Garcia-Luque, E., Gašparović, B., Godrijan, J., Lawlor, P.J., Maes, F., Malej, A., Panagiotaras, D., 2020. European policies and legislation targeting ocean acidification in european waters-current state. Mar. Policy 118, 103947.

Galland, G., Harrould-Kolieb, E., Herr, D., 2012. The ocean and climate change policy. Clim Policy. 12, 764–771.

Groeneveld, R.A., 2020. Welfare economics and wicked problems in coastal and marine governance. Mar. Policy 117, 103945.

Guggisberg, S., 2019. Funding coastal and marine fisheries projects under the climate change regime. Mar. Policy 107, 103352.

Güssow, K., Proelss, A., Oschlies, A., Rehdanz, K., Rickels, W., 2010. Ocean iron fertilization: why further research is needed. Mar. Policy 34, 911–918.

Hanich, Q., Wabnitz, C.C., Ota, Y., Amos, M., Donato-Hunt, C., Hunt, A., 2018. Small-scale fisheries under climate change in the Pacific Islands region. Mar. Policy 88, 279–284.

Harrould-Kolieb, E.R., 2019. (re) Framing Ocean acidification in the context of the United Nations framework convention on climate change (UNFCCC) and Paris agreement. Clim Policy. 19, 1225–1238.

Hoel, A.H., 2020. Oceans, fish and resource Management in the Northern Areas. Int Polit-Oslo. 78, 500–510.

Hu, C.P., Hu, J.M., Gao, Y., Zhang, Y.K., 2011. A journal co-citation analysis of library and information science in China. Scientometrics. 86, 657–670.

Johnson, D., Ferreira, M.A., Kenchington, E., 2018. Climate change is likely to severely limit the effectiveness of deep-sea ABMTs in the North Atlantic. Mar. Policy 87, 111–122.

Jones, P.J., 2021. A governance analysis of Ningaloo and Shark Bay marine parks, Western Australia: putting the 'eco'in tourism to build resilience but threatened in long-term by climate change? Mar. Policy 127, 103636.

Kaiser, B.A., Bakanev, S., Bertelsen, R.G., Carson, M., Eide, A., Fernandez, L., Halpin, P., Izmalkov, S., Kyhn, L.A., Österblom, H., Punt, M., 2015. Spatial issues in Arctic marine resource governance workshop summary and comment. Mar. Policy 58, 1–5.

Khan, A., Charles, A., Armitage, D., 2018. Place-based or sector-based adaptation? A case study of municipal and fishery policy integration. Clim Policy. 18, 14–23.

Le Cornu, E., Doerr, A.N., Finkbeiner, E.M., Gourlie, D., Crowder, L.B., 2018. Spatial management in small-scale fisheries: a potential approach for climate change adaptation in Pacific Islands. Mar. Policy 88, 350–358. https://doi.org/10.1016/j.marpol.2017.09.030.

Lim, M., 2016. Governance criteria for effective transboundary biodiversity conservation. Int. Environ. Agreements 16, 797–813. https://doi.org/10.1007/s10784-015-9296-3.

Macpherson, E., Urlich, S.C., Rennie, H.G., Paul, A., Fisher, K., Braid, L., Banwell, J., Ventura, J.T., Jorgensen, E., 2021. 'Hooks' and 'anchors' for relational ecosystem-based marine management. Mar. Policy 130, 104561.

Malone, T.C., DiGiacomo, P.M., Gonçalves, E., Knap, A.H., Talaue-McManus, L., de Mora, S., 2014. A global ocean observing system framework for sustainable development. Mar. Policy 43, 262–272.

McCain, K.W., 1991. Mapping economics through the journal literature: an experiment in journal cocitation analysis. J. Am. Soc. Inf. Sci. 42, 290–296.

Mursyid, H., Daulay, M.H., Pratama, A.A., Laraswati, D., Novita, N., Malik, A., Maryudi, A., 2021. Governance issues related to the management and conservation of mangrove ecosystems to support climate change mitigation actions in Indonesia. Forest Policy Econ. 133, 102622.

Nagarajan, P., 2022. Stockholm +50 Seeks a Healthy Planet for the Prosperity of all. Available from https://www.saltwire.com/atlantic-canada/opinion/commentary-stockholm-50-seeks-a-healthy-planet-for-the-prosperity-of-all-100742261/. (Accessed 16 Aug. 2022).

Niiranen, S., Richter, A., Blenckner, T., Stige, L.C., Valman, M., Eikeset, A.M., 2018. Global connectivity and cross-scale interactions create uncertainty for blue growth of Arctic fisheries. Mar. Policy 87, 321–330. https://doi.org/10.1016/j.marpol.2017.10.024.

Nilsson, J.A., Fulton, E.A., Haward, M., Johnson, C., 2016. Consensus management in Antarctica's high seas–past success and current challenges. Mar. Policy 73, 172–180.

Nursey-Bray, M., Magnusson, A., Bicknell, N., Magnusson, M., Morison, J., Sullivan, A., 2018. Adapting to change: Prioritising management for the future of the marine Scalefish fishery. Mar. Policy 95, 153–165.

Ogier, E.M., Davidson, J., Fidelman, P., Haward, M., Hobday, A.J., Holbrook, N.J., Hoshino, E., Pecl, G.T., 2016. Fisheries management approaches as platforms for climate change adaptation: comparing theory and practice in Australian fisheries. Mar. Policy 71, 82–93.

Österblom, H., Gårdmark, A., Bergström, L., Müller-Karulis, B., Folke, C., Lindegren, M., Casini, M., Olsson, P., Diekmann, R., Blenckner, T., Humborg, C., 2010. Making the ecosystem approach operational—can regime shifts in ecological-and governance systems facilitate the transition? Mar. Policy 34, 1290–1299.

Pentz, B., Klenk, N., Ogle, S., Fisher, J.A., 2018. Can regional fisheries management organizations (RFMOs) manage resources effectively during climate change? Mar. Policy 92, 13–20.

Pittman, J., Armitage, D., 2019. Network governance of land-sea social-ecological systems in the Lesser Antilles. Ecol. Econ. 157, 61–70. https://doi.org/10.1016/j.ecolecon.2018.10.013.

Porobic, J., Fulton, E.A., Frusher, S., Parada, C., Haward, M., Ernst, B., Stram, D., 2018. Implementing ecosystem-based fisheries management: lessons from Chile's experience. Mar. Policy 97, 82–90.

Rahman, H.T., Bowron, T., Pett, B., Sherren, K., Wilson, A., van Proosdij, D., 2021. Navigating nature-based coastal adaptation through barriers: a synthesis of practitioners' narrative from Nova Scotia, Canada. Soc Nat. Resour. 34, 1268–1285.

Ricart, S., Rico-Amorós, A.M., 2022. Can agriculture and conservation be compatible in a coastal wetland? Balancing stakeholders' narratives and interactions in the management of El Hondo Natural Park, Spain. Agr Hum Values. 39, 589–604. https://doi.org/10.1007/s10460-021-10271-5.

Ringbom, H., Joas, M., 2018. Concluding article: the changing regulatory landscape of the Baltic Sea–an analysis. Mar. Policy 98, 317–324. https://doi.org/10.1016/j.marpol.2018.09.029.

Rogers, K., Boon, P.I., Branigan, S., Duke, N.C., Field, C.D., Fitzsimons, J.A., Kirkman, H., Mackenzie, J.R., Saintilan, N., 2016. The state of legislation and policy protecting Australia's mangrove and salt marsh and their ecosystem services. Mar. Policy 72, 139–155.

Saarman, E.T., Carr, M.H., 2013. The California marine life protection act: a balance of top down and bottom up governance in MPA planning. Mar. Policy 41, 41–49. https://doi.org/10.1016/j.marpol.2013.01.004.

Scharin, H., Ericsdotter, S., Elliott, M., Turner, R.K., Niiranen, S., Blenckner, T., Hyytiäinen, K., Ahlvik, L., Ahtiainen, H., Artell, J., Hasselström, L., 2016. Processes for the sustainable stewardship of marine environments. Ecol. Econ. 128, 55–67.

Serrao-Neumann, S., Davidson, J.L., Baldwin, C.L., Dedekorkut-Howes, A., Ellison, J.C., Holbrook, N.J., Howes, M., Jacobson, C., Morgan, E.A., 2016. Marine governance to avoid tipping points: can we adapt the adaptability envelope? Mar. Policy 65, 56–67.

Shiiba, N., Wu, H.H., Huang, M.C., Tanaka, H., 2022. How blue financing can sustain ocean conservation and development: a proposed conceptual framework for blue financing mechanism. Mar. Policy 139, 104575.

Sumaila, U.R., Bellmann, C., Tipping, A., 2016. Fishing for the future: an overview of challenges and opportunities. Mar. Policy 69, 173–180.

Sumby, J., Haward, M., Fulton, E.A., Pecl, G.T., 2021. Hot fish: the response to climate change by regional fisheries bodies. Mar. Policy 123, 104284.

Talloni-Álvarez, N.E., Sumaila, U.R., Le Billon, P., Cheung, W.W., 2019. Climate change impact on Canada's Pacific marine ecosystem: the current state of knowledge. Mar. Policy 104, 163–176.

Teneva, L., Strong, A.L., Agostini, V., Bagstad, K.J., Drakou, E.G., Ancona, Z., Gjerde, K., Hume, A.C., Jickling, N., 2022. Estimating the pelagic ocean's benefits to humanity can Enhance Ocean Governancehumanity can enhance science diplomacy and ocean governance. Mar. Policy 136, 104906. https://doi.org/10.1016/j.marpol.2021.104906.

Truchet, D.M., Noceti, B.M., Villagran, D.M., Truchet, R.M., 2022. Alternative conservation paradigms and ecological knowledge of small-scale artisanal fishers in a changing marine scenario in Argentina. Hum. Ecol. 50, 209–225.

UN, 2022. Guterres outlines four recommendations to help us all 'Save Our Ocean'. Available from https://news.un.org/en/story/2022/06/1121402. Accessed 16 Aug. 2022.

Văidianu, N., Tătui, F., Ristea, M., Stănică, A., 2020. Managing coastal protection through multi-scale governance structures in Romania. Mar. Policy 112, 103567.

Van der Geest, C., 2017. Redesigning Indian Ocean fisheries governance for 21st century sustainability. Glob Policy. 8, 227–236. https://doi.org/10.1111/1758-5899.12447.

Villarrubia-Gómez, P., Cornell, S.E., Fabres, J., 2018. Marine plastic pollution as a planetary boundary threat–the drifting piece in the sustainability puzzle. Mar. Policy 96, 213–220.

Villasante, S., Rodríguez, D., Antelo, M., Quaas, M., Österblom, H., 2012. The global seafood market performance index: a theoretical proposal and potential empirical applications. Mar. Policy 36, 142–152.

Warner, R., 2018. Oceans in transition: incorporating climate-change impacts into environmental impact assessment for marine areas beyond national jurisdiction. Ecol. Law Quart. 45, 31–52.

Wei, F., Grubesic, T.H., Bishop, B.W., 2015. Exploring the GIS knowledge domain using CiteSpace. Prof Geogr. 67, 374–384. https://doi.org/10.1080/00330124.2014.983588.

Wendebourg, M.R., 2020. Southern Ocean fishery management-is CCAMLR addressing the challenges posed by a changing climate? Mar. Policy 118, 103847.

Wood, A.L., Butler, J.R., Sheaves, M., Wani, J., 2013. Sport fisheries: opportunities and challenges for diversifying coastal livelihoods in the Pacific. Mar. Policy 42, 305–314.

WWF, 2015. Reviving the Ocean Economy: The Case for Action. World Wildlife Fund, p. 7.

Yadav, S.S., Gjerde, K.M., 2020. The ocean, climate change and resilience: making ocean areas beyond national jurisdiction more resilient to climate change and other anthropogenic activities. Mar. Policy 122, 104184. https://doi.org/10.1016/j.marpol.2020.104184.

Yates, K.L., Payo, A.P., Schoeman, D.S., 2013. International, regional and national commitments meet local implementation: a case study of marine conservation in Northern Ireland. Mar. Policy 38, 140–150.

Yu, D., Xu, Z., Pedrycz, W., Wang, W., 2017. Information sciences 1968–2016: a retrospective analysis with text mining and bibliometric. Inform. Sci. 48, 619–634.

FSC
www.fsc.org
MIX
Papier | Fördert gute Waldnutzung
FSC® C083411

Kolibri 360 GmbH
Ferdinand-Jühlke-Straße 7
99095 Erfurt, Deutschland
produktsicherheit@kolibri360.de